Memoirs of the American Mathematical Society
Number 146

Sandy Grabiner

Derivations and Automorphisms of Banach Algebras of Power Series

Published by the
American Mathematical Society
Providence, Rhode Island
1974

AMS (MOS) 1970 Subject Classifications: 30A98, 46H05, 46J15, 46J99, 47A15, 47A60, 47B99, 47C05; 30A78, 45D05, 46L20.

Library of Congress Cataloging in Publication Data **CIP**

Grabiner, Sandy, 1939–

Derivations and automorphisms of Banach algebras of power series.

(Memoirs of the American Mathematical Society, no. 146)

Bibliography: p.

1. Banach algebras. 2. Power series. 3. Linear operators. I. Title. II. Series: American Mathematical Society. Memoirs, no. 146.

QA3.A57 no. 146 [QA326] 510'.8s [512'.55] 74-7124

ISBN 0-8218-1846-5

CONTENTS

PART I. INTRODUCTION

1. General Introduction 1

2. Statements of Major Results 6

3. Multiplying and Dividing Power Series 14

PART II. GENERAL THEORY

4. Derivations 20

5. Endomorphisms 26

6. Automorphisms 30

7. The Group of Automorphisms 43

8. Exponentials of Derivations 49

9. Derivations and Automorphisms of Ideals 56

10. Rates of Quasi-Nilpotence 65

11. Orbits of the Automorphism Group 74

PART III. APPLICATIONS

12. Weighted Shifts 80

13. Spaces of Analytic Functions 94

14. Operational Calculus 114

Abstract

This paper studies derivations, endomorphisms, automorphisms, and various related questions about certain Banach algebras, B, which are continuously embedded in the space of complex formal power series in the indeterminate z. It is shown, under appropriate hypotheses on $\{\|z^n\|\}$, that if $g = \Sigma_2^\infty \lambda_n z^n$ belongs to B, then there is a unique derivation of B which maps z to g; there is a unique endomorphism that maps z to g; and there is a unique automorphism that maps z to $z + g$. The structure of the automorphism group of B, as a topological group in the uniform operator topology, is described. Applications are given to the study of weighted shifts and their commuting algebras, to the study of H^p spaces and similar spaces of analytic functions, and to the study of quasi-nilpotent operators.

BANACH ALGEBRAS OF POWER SERIES

1. INTRODUCTION

In this paper we study the derivations, automorphisms, and endomorphisms of a class of radical Banach algebras. The algebras we consider will all be subalgebras of $C[[z]]$, the algebra of complex formal power series in one indeterminate. Explicitly, we will consider certain algebras of the type described in the following definition.

DEFINITION (1.1). Suppose that B is an algebra of formal power series with zero constant term, and is a Banach space under some norm. B is a *generalized Banach algebra of power series*, if it contains the indeterminate z and if all the *coefficient projections* $\Sigma \lambda_n z^n \to \lambda_k$ are continuous. B is a *Banach algebra of power series* if, in addition, the polynomials are dense in B. If B is a generalized Banach algebra of power series, then $B^{\#}$ is the Banach space sum of B and the constant power series.

Because of the closed graph theorem, the power series multiplication on B and $B^{\#}$ is separately continuous. So, if B is a generalized Banach algebra of power series, then B and $B^{\#}$ have equivalent norms under which they are in fact Banach algebras [38, p. 5]. Our use of the terms (generalized) Banach algebra of power series corresponds to the use in [15]. Slightly differing terminologies are used by Lorch and Shilov [26] [41] [48, p.317] and by Loy [27] [28] [29] [30].

In Sections 4, 5, and 6, respectively, we will give explicit formulas for essentially all the derivations, endomorphisms, and automorphisms of the algebras we consider. In these sections and in the rest of Part II, we consider related questions such as: the spectra of derivations, endomorphisms,

---

Received by the Editors September 25, 1972; and in revised form September 15, 1973. Revised version supported by NSF Grant NO. GP-37817.

and automorphisms; the structure of the group of automorphisms and of its orbits; and the relation between derivations and automorphisms. A detailed statement of our major results can be found in Section 2, particularly in Theorem (2.6). The characterization of automorphisms, in Section 6, is the major result in this paper, and can be read independently of Sections 4 and 5.

The simplest Banach spaces of power series are those in which the underlying Banach space is some l_p. Since we will frequently want to specialize our results to these spaces, we repeat the following definition from [15, Def. (6.1)].

DEFINITION (1.2). Suppose that $\{c_n\}_1^\infty$ is a sequence of non-zero scalars for which $\{c_{n+1}/c_n\}$ is bounded, and suppose that $1 \le p \le \infty$. $K^p\langle c_n\rangle$ is the Banach space of all series $f = \Sigma_1^\infty \lambda_n z^n$ for which $\{\lambda_n c_n\}$ belongs to l_p. The norm of f in $K^p\langle c_n\rangle$ is the l_p norm of $\{\lambda_n c_n\}$. We will usually write $K\langle c_n\rangle$ for $K^1\langle c_n\rangle$; and, when $\{c_n\}$ is understood, we will often write K^p and K for $K^p\langle c_n\rangle$ and $K\langle c_n\rangle$, respectively.

There are a variety of sufficient conditions on $\{c_n\}$ which guarantee that $K^p\langle c_n\rangle$ is a radical algebra (see [15, Section 6]). The simplest sufficient condition is that there exists a fixed integer k for which $\{c_{n+k}/c_n\}_n$ is eventually non-increasing and for which $\Sigma_n|c_{n+k}|/|c_n|$ converges (see [13, Thm. (2.10)(A), p. 645] and [13, Lemma (3.6), p. 648]).

Notice that the operation of multiplication by z on $K^p\langle c_n\rangle$ corresponds to the weighted shift with weights $\{c_{n+1}/c_n\}$ on l_p. In fact, one of the major applications of the theory of Banach algebras of power series has been to the study of weighted shifts on l_p and on more general spaces ([36][37][6][7][15]). In Section 12 below, we apply the results of the present paper to the study of weighted shifts. Among other things, we determine operators similar to a given weighted shift L, and we characterize those operators which generate the same algebra as L. We will also determine all closed operators which commute with L (this generalizes [15, Thm. (5.1)]). The statements of our major results about weighted shifts do not involve power series, and can be read independently of the rest of this paper. Indeed, most of Section 12 can be read after Section 2.

Another context in which Banach algebras of power series arise is in the study of convolutions of analytic functions. If $F(x)$ and $G(x)$ are analytic functions of x on some region D which is star-shaped about the origin, then we define the convolution product:

$$F*G(x) = \int_0^x F(x-t)G(t)dt.$$

We can associate with each function $F(x)$ analytic on D a formal power series $\hat{F}(z)$ in such a way that $\hat{F}\hat{G}$ is associated with $F*G$ (see Section 13, below, for the details). Although such spaces of analytic functions provided one of the earliest examples of radical Banach algebras of power series [41, pp. 103 and 114], I know of only two short papers [8] [11] in which the power series representation of convolution has been exploited. Section 13 is devoted to applying the methods and results of the present paper, and of our earlier papers [13] [14] [15], to the study of the structure of spaces of analytic function on the disc, such as the H^p spaces, under convolution. We also study the anti-derivative operator, $1*F$, and we discuss the solution of certain integral equations in these spaces.

The statements of the major results in Section 13 can be read independently of the rest of this paper; and most of the other parts of Section 13 can be read after Section 2.

Many of the algebras of analytic functions that we study in Section 13 are subalgebras of $L^1(0,1)$. We will frequently point out similarities and differences between our results on the derivations and automorphisms of Banach algebras of power series and the results proved by Scheinberg and Kamowitz [21] for $L^1(0,1)$. We will also point out open problems about $L^1(0,1)$ which are suggested by our results.

Radical algebras of power series have proved useful in studying arbitrary quasi-nilpotent operators [12] [13] [14] [16, Section 2]. Given a quasi-nilpotent operator T and a fixed operator topology, one considers the collection of formal power series $f = \Sigma_1^\infty \lambda_n z^n$ for which the operator $f(T) = \Sigma\lambda_n T^n$ is defined. In Section 14, we use the results of Part II to investigate the relation between T and $f(T)$, and we determine conditions under which $g \circ f(T) = g(f(T))$ (extending the results in [13] in which g was assumed to have positive radius of convergence).

Radical Banach algebras of power series have been studied in their own right largely because they provide the simplest examples of radical Banach algebras, and hence can serve as one starting point for a general theory of radical Banach algebras. In particular, several papers have considered questions related to derivations, automorphisms, and endomorphisms of (generalized) Banach algebras of power series. Loy [27] [28] [29] [30] has shown that if B is a generalized Banach algebra of power series, then all derivations, automorphisms, and endomorphisms of B and $B^{\#}$ are automatically continuous. Ginsberg and Newman [8, Th. 1] describe the generators of certain $K\langle c_n\rangle$; we will see, in Section 6 below, that their results can be obtained as a consequence of our study of the automorphisms of these algebras. Newman [34] and Scheinberg [40, p. 322] have constructed examples of radical $K<c_n>$ with no non-zero derivations; and Scheinberg [40, Th. 5, p. 322] has shown that these algebras have only the most obvious automorphisms. While we do include generalizations of these results to algebras other than $K<c_n>$, we emphasize the opposite extreme, the case in which the Banach algebras have the maximum number of derivations, automorphisms, and endomorphisms. Building on Loy's results [27], and the results in the present paper, Schultz [42] has shown that systems of derivations on Banach algebras of power series are automatically continuous, and he has obtained explicit formulas for essentially all systems of derivations of the algebras studied in this paper.

2. STATEMENT OF MAJOR RESULTS

In this section, we state the major results about generalized Banach algebras of power series proved in this paper; and we briefly outline the contents of the various sections of the paper. We begin with some basic definitions which introduce the notation and terminology that we will use in the statements and proofs of our theorems about derivations, endomorphisms, and automorphisms.

DEFINITION (2.1). Suppose that g is a power series, that h is a power series with zero constant term, and that k is a non-negative integer. We define the operators L_g, D, gD, and T_h on $C[[z]]$ and the linear functional π_k on $C[[z]]$ by:

(A) $L_g f = gf$.

(B) $D(\Sigma_0^\infty \lambda_n z^n) = \Sigma_1^\infty n\lambda_n z^{n-1}$.

(C) $gD = L_g D$.

(D) $T_h f = f \circ h$ (formal composition of power series).

(E) $\pi_k (\Sigma_0^\infty \lambda_n z^n) = \lambda_k$.

DEFINITION (2.2). Suppose that f is a formal power series and that n is a non-negative integer. We say f has *order* n if $\pi_n(f) \neq 0$, but $\pi_k(f) = 0$ for all $k < n$ (if $f = 0$, f has *order* ∞). If B is a generalized Banach algebra of power series, $B^{(n)}$ is the collection of all series in $B^\#$ of order at least n.

From now on, we will assume that $C[[z]]$ is given the weak topology determined by the coefficient projections $\{\pi_n\}$ (see [45, p. 28] or [39, pp. 323-324]). This topology on $C[[z]]$, or on any of its subsets, will be called the *Fréchet* topology. The maps L_g, gD, and T_h are clearly all continuous in the Fréchet topology. Therefore, their restrictions to maps between Banach spaces continuously embedded in $C[[z]]$ (such as $B^{(n)}$, for B a generalized Banach algebra of power series) are closed linear transformations.

It is well known that T_h is always an automorphism of $C[[z]]$ and that gD is always a derivation [2, Chapter 1] [39]. Consequently, to say that T_h is an endomorphism (automorphism) of a generalized Banach algebra of power series B, is the same as saying that T_h maps B into (onto) itself; and similarly for derivations. If B is actually a Banach algebra of power series (that is, if the polynomials are dense in B), it is not hard to show that every endomorphism of B must be of the form T_h for some h in B, and that every derivation is of the form gD for some g in B (this is implicit in [40]). In Theorem (2.7) at the end of this section, we extend this result to show that $B^{\#}$ also has the same set of derivations, endomorphisms, and automorphisms as B; and we also consider certain cases in which B is only a generalized Banach algebra of power series. Thus the problem of determining all derivations, endomorphisms, and automorphisms of B and $B^{\#}$ is reduced to the problem of determining which gD map B into itself, and which T_h map B into or onto itself.

Those automorphisms T_h of $B^{\#}$ for which h belongs to $z + B^{(2)}$ will play a special role in this paper, so the following definition will be convenient.

DEFINITION (2.3). Suppose that B is a generalized Banach algebra of power series. An automorphism T of $B^{\#}$ will be called *principal* if Tz belongs to $z + B^{(2)}$ and will be called *scalar* if Tz is a scalar multiple of z.

We now define a collection of power series, which will play a crucial role in this paper, just as they did in [13], [14], and [15].

DEFINITION (2.4). Suppose that B is a generalized Banach algebra of power series and that j is a non-negative integer. $S_{-j}(B)$ is the Banach space of those formal power series f with zero constant term for which fz^j belongs to B. The norm of f in S_{-j} is set equal to the norm of fz^j in B. When the algebra B is clearly understood, we will often write S_{-j} instead of $S_{-j}(B)$.

The hypothesis of most of our theorems will involve assumptions that some or all $S_{-j}(B)$ are algebras or radical algebras. In the next section we will mention several sufficient conditions for this. Often it will be enough to show that all $S_{-j}(K<c_n>)$ are algebras, where c_n is the norm of z^n in B, and this follows, for instance, whenever there is a k for which $\{c_{n+k}/c_n\}_n$ is eventually non-increasing [13, Th. (2.10), p. 645].

We need one final definition before stating our major results.

DEFINITION (2.5). If X is a Banach space continuously embedded in $C[[z]]$ and if f is the formal power series $\Sigma\lambda_n z^n$, we say that f is *convergent in* X if the series $\Sigma\lambda_n z^n$ converges to f in the norm topology on X.

We are now ready to state the major results that we will prove about derivations, endomorphisms, and automorphisms of generalized Banach algebras of power series. In the theorem below, we choose a single set of hypotheses strong enough to imply all of the results. However, most of the separate parts of the theorem will be proved under weaker hypotheses, and in somewhat more generality, in Sections 4 through 10. For the special case $B = K^p<c_n>$, $1 \le p < \infty$; hypothesis (ii) of the following theorem simply says that $\{nc_{n+1}/c_n\}$ is bounded.

THEOREM (2.6). *Suppose that* B *is a radical generalized Banach algebra of power series and that* $c_n = \|z^n\|$. *If*

(i) $S_{-j}(K<c_n>)$ *is an algebra for all non-negative integers* j,

(ii) z^2Df *is convergent in* B *for all* f *in* $B^\#$;

Then:

(A) gD *is a derivation of* $B^\#$ *if and only if* g *belongs to* $B^{(2)}$.

(B) *The map* $g \to gD$ *is a linear homeomorphism from* $B^{(2)}$ *onto the set of derivations of* $B^\#$ *with the uniform operator topology.*

(C) T_g *is a compact endomorphism of* $B^\#$ *for all* g *in* $B^{(2)}$.

(D) *If* T_g *is an endomorphism of* $B^\#$, *then* $|\pi_1(g)| \leq 1$. *If* T_g *is an automorphism, then* $|\pi_1(g)| = 1$.

(E) T_g *is a principal automorphism of* $B^\#$ *if and only if* g *belongs to* $z + B^{(2)}$.

(F) $z + B^{(2)}$ *is a topological group, in the relative norm topology, under the operation of formal power series composition.*

(G) *The map* $g \to T_g$ *is a homeomorphic group anti-isomorphism from* $z + B^{(2)}$ *onto the group of principal automorphisms of* $B^\#$.

(H) *The automorphism group of* $B^\#$, *in the uniform operator topology, is algebraically and topologically a semi-direct product of the principal automorphism subgroup with the scalar automorphism subgroup. The principal automorphism subgroup is the component of the identity, and the scalar automorphism subgroup is discrete.*

(I) *The principal automorphisms of* $B^\#$ *are precisely the set of automorphisms which are the product of two exponentials of derivations.*

(J) *If* g *belongs to* $z^k + B^{(k+1)}$, *then there is an automorphism*

of $B^{(k)}$ *which maps* z^k *to* g.

(K) *If* f^k *belongs to* $z^k + B^{(k+1)}$, *then the sequence* $\{\|f^n\|/\|z^n\|\}_{n=k}^{\infty}$ *is bounded above and below.*

(L) V *is a closed operator which commutes with* L_z *if and only if either* V *or* V^{-1} *equals* L_f *for a series* f *in some* $S_{-j}^{\#}$; *moreover, all such* L_f *are continuous operators on* $B^{(j)}$.

(M) *If* $\{z^n\}$ *is a Schauder basis of* $B^{\#}$, *then* T_g *is a compact endomorphism of* $B^{\#}$ *if and only if* $|\pi_1(g)| < 1$.

(N) *If* $\{z^n\}$ *is an unconditional basis of* $B^{\#}$, *then* T_g *is an automorphism of* $B^{\#}$ *if and only if* $|\pi_1(g)| = 1$.

In Section 3, we will discuss when the hypotheses of Theorem (2.6), above, and its generalizations are satisfied. We will also derive some lemmas about products and quotients of power series in B, under the assumption that $S_{-j}(B)$ is an algebra or a radical algebra. The results in Section 3 are all either repeated without proof or else slightly adapted from results in our earlier papers [13] [14] [15].

Section 4 on derivations and Section 5 on endomorphisms are the most elementary sections of this paper. The arguments in these sections involve only the results of Section 3, and are quite similar to arguments in [13] [14] [15]. These sections include proofs of (2.6) (A), (B), (C), (D), and (M).

Section 6, which is mainly devoted to establishing the characterization of principal automorphisms given in (2.6) (E), is the deepest and most important section of this paper. The results in Section 6 are based on techniques quite different from those we used in Sections 4 and 5 and in our earlier papers [13] [14] [15]. The techniques and results of Section 6 are the basic tools for the rest of Part II; and they are used in most of the applications to weighted shifts and to the operational calculus for quasi-nilpotent operators in Sections 12 and 14, respectively.

Section 7 studies the group of automorphisms of $B^{\#}$, the structure of $z + B^{(2)}$ under power series composition, and the non-principal automorphisms of $B^{\#}$. It includes proofs of (2.7) (F), (G), (H), and (N).

In Section 8, we combine the results on automorphisms in Section 6 with known results on the exponentials of derivations in general Banach algebras (from [22], for instance) to study the map $gD \to \exp(gD)$, which maps the derivations of $B^{\#}$ into the principal automorphisms. We prove (2.7) (I) and apply it to determine the spectra of principal automorphisms. We also show that, if z^2D is also compact, the map $gD \to \exp(gD)$ is a homeomorphism from the derivations of $B^{\#}$ onto the principal automorphisms.

In Section 9, we characterize the derivations, endomorphisms, and automorphisms of $B^{(k)}$, proving (2.6) (J) in particular. These results are applied in Section 10 to obtain estimates of the norms of powers, and of products, of elements of B. Theorem (2.6) (K) is typical of these estimates. These estimates are important in our applications to the operational calculus, in Section 14.

Section 11 contains some preliminary results on the structure of the orbits of the principal automorphism group. This section completes Part II, on the general theory of Banach algebras of power series. The proof of Theorem (2.7) (L), which describes the closed operators commuting with the shift L_z, is given in Section 12, on weighted shifts.

The following theorem shows that all derivations and endomorphisms of $B^{\#}$ are of the form gD and T_g, respectively, for some g in B. Together with Theorem (2.6), above, this provides an explicit determination of the derivations, endomorphisms, and automorphisms of the algebras we study. One important consequence of the following theorem is that B and $B^{\#}$ have precisely the same derivations, endomorphisms, and automorphisms. The hypothesis on B in the following theorem is clearly satisfied if B is a radical Banach algebra of power series. We will see, in Lemma (3.1), below, that it is also satisfied if B is a generalized Banach algebra of power series for which z^2Df converges in B for all f in $B^{\#}$.

THEOREM (2.7). *Suppose that* B *is a radical generalized Banach algebra of power series, that* R *is the closure of the polynomials in* B, *and that* V *is a linear transformation on* $B^{\#}$ *with* $Vz = g$. *If there is a non-negative integer* k *for which* $Bz^k \subseteq R$, *then:*

(A) V *is a derivation of* $B^{\#}$ *if and only if* g *belongs to* B *and* V *is the restriction of* gD *to* $B^{\#}$.

(B) V *is an endomorphism of* $B^{\#}$ *if and only if* g *belongs to* B *and* V *is the restriction of* T_g *to* $B^{\#}$.

Proof. We already know that V is a derivation if it equals gD and is an endomorphism if it equals T_g. If V is a derivation it is automatically continuous (see [27, Th. 1, p. 420] or [28, Th. 2.1, p. 373]), and similarly if V is an endomorphism (see [30, Th. 10] or [20, Th. 9.1, p. 99]).

Suppose now that V is a derivation of $B^{\#}$ with $Vz = g$. Then, for all non-negative integers n, we have

$$V(z^n) = nz^{n-1}g. \tag{2.8}$$

If g did not belong to B, then it would be invertible in $B^{\#}$, since B is radical [38, Th. (2.3.2)(iii), p. 55]. Then the continuous map

$f \to zg^{-1}V(f)$ would take z^n to nz^n. Since this is impossible, g belongs to B. By formula (2.8) the closed linear map gD agrees with V on the polynomials, so V and gD agree on $R^{\#}$. Suppose now that f is an arbitrary series in $B^{\#}$. Since fz^k belongs to $R^{\#}$, we have

$$fV(z^k) + V(f)z^k = V(fz^k) = gD(fz^k) = fgD(z^k) + g(Df)z^k.$$

Since $V(z^k) = gD(z^k)$, we have $V(f)z^k = (gDf)z^k$; so $V = gD$ throughout $B^{\#}$.

Similarly, suppose that V is an endomorphism of $B^{\#}$ with $Vz = g$. Since z is quasi-nilpotent and V is continuous, g is quasi-nilpotent and must therefore belong to B, the radical of $B^{\#}$ [38, Cor. (2.3.6), p. 57]. As above, it is easy to show that V and T_g agree on $R^{\#}$. Suppose, therefore, that f is an arbitrary series in $B^{\#}$. Then

$$(T_g f)g^k = T_g(fz^k) = V(fz^k) = V(f)V(z^k) = V(f)g^k.$$

So $T_g \equiv V$, and the proof is complete.

3. MULTIPLYING AND DIVIDING POWER SERIES

Suppose that B is a generalized Banach algebra of power series, and let $\|z^n\| = c_n$. In this section we examine the relation between various assumptions on $\{c_n\}$, $z^m Df$, $S_{-j}(B)$, and $S_{-j}(K)$. These results will be useful in determining when B satisfies the hypotheses of the various theorems we prove in this paper, and also in relating the hypotheses of different theorems. In particular, the results in this section show that the hypotheses we use in later sections to prove the various parts of Theorem (2.6) are all in fact weaker than those given in Theorem (2.6).

In this section we also prove several results about products, quotients, powers, and roots of power series, under the hypothesis that some $S_{-j}(B)$ is an algebra or a radical algebra. These results will be used frequently in this paper.

The results in this section are lemmas useful for proving our major results on derivations, endomorphisms, and automorphisms. Most of these results are repetitions of, or else slightly adapted from, results in [13], [14], and [15].

The case $B = K\langle c_n\rangle$ is particularly simple. If j is a fixed integer and $d_n = c_{n+j}$, then $S_{-j}(K\langle c_n\rangle)$ is just $K\langle d_n\rangle$. So $S_{-j}(K)$ is a radical algebra precisely when $\lim(c_n)^{1/n} = 0$ and there is an $M > 0$ for which $d_{n+m} \leq M d_n d_m$, for all n, m. (See [13, Lemma (2.8), p. 645]). In principle, this completely solves the question of when $S_{-j}(K\langle c_n\rangle)$ is a radical algebra, but these conditions on $\{c_n\}$ are not always easy to verify. The simplest sufficient condition on $\{c_n\}$ for all $S_{-j}(K\langle c_n\rangle)$ to be radical is that there exist an integer k for which $\{c_{n+k}/c_n\}_n$ is eventually non-increasing and has limit 0 (see [13, Th. (2.10), p. 645]). For other such sufficient conditions on $\{c_n\}$, see the discussion after [15, Lemma (3.3)].

Sufficient conditions on a general B for $S_{-j}(B)$ to be an algebra or a radical algebra are usually much more complicated (see for instance [15, Th. (6.7)]). The next two lemmas, taken together, show that in many important cases it is enough to show that all $S_{-j}(K\langle c_n\rangle)$ are algebras.

LEMMA (3.1). *Suppose that* B *is a generalized Banach algebra of power series with* $\|z^n\| = c_n$. *If there is a positive integer* k *such that* $z^{k+1}Df$ *converges in* B *for all* f *in* B, *then* $\lim(c_n)^{1/n} = 0$, *and there is an integer* m *for which* $B \subseteq S_{-m}(K\langle c_n\rangle)$.

Proof. Since $z^{k+1}Dz^n = nz^{k+n}$, we have

$$c_{n+k} \le c_k\|z^{k+1}D\|/n. \tag{3.2}$$

Therefore $\lim_n(c_{n+k}/c_k) = 0$, so z^k is quasi-nilpotent in B [1, Prob. 12-4, p. 383]. Hence z is also quasi-nilpotent, which means $\lim(c_n)^{1/n} = 0$.

Also, for each f in B, the series $z^{k+1}Df = \Sigma_n\, n\, \pi_n(f)z^{n+k}$ converges in B. Hence, by the Banach-Steinhaus uniform boundedness theorem, there is an $M > 0$ for which

$$\|\pi_n\|c_{n+k} \le M/n. \tag{3.3}$$

Combining formulas (3.2) and (3.3), we obtain

$$\|\pi_n\|c_{n+2k} \le \frac{\|z^{k+1}D\|\ \|\pi_n\|c_{n+k}}{n+k} \le \frac{M\|z^{k+1}D\|}{n(n+k)}.$$

Hence $\Sigma_n\|\pi_n\|c_{n+k}$ converges, so $B \subseteq S_{-2k}(K\langle c_n\rangle)$. This completes the proof.

LEMMA (3.4). *Suppose* B *is a generalized Banach algebra of power series with* $\|z_n\| = c_n$. *Suppose also that* $\lim(c_n)^{1/n} = 0$, $B \subseteq S_{-m}(K\langle c_n\rangle)$ *for some* m, *and* $S_{-j}(K)$ *is an algebra for all* j. *Then all* $S_{-j}(B)$ *are radical algebras.*

Proof. Fix j. Then $zS_{-j}(B) \subseteq S_{-j}(B)$ and $K\langle c_n\rangle \subseteq S_{-j}(B) \subseteq S_{-(j+m)}(K)$. So $S_{-j}(B)$ is an algebra by [13, Lemma (3.6), p. 648]. $S_{-j}(B)$ is a radical algebra by [13, Lemma (3.8), p. 649] or by Lemma (3.6) below.

In most of our results we will only assume that some specific $S_{-j}(B)$ is an algebra or a radical algebra. We do this because there are examples for which only some $S_{-j}(B)$ are (radical) algebras [9, pp. 136-137]. The reader may, however, find it easier to assume that all $S_{-j}(B)$ are radical algebras, and thus not have to keep track of the varying hypotheses in different results. The next two results compare conditions on differing $S_{-j}(B)$.

LEMMA (3.5). *Suppose that* B *is a generalized Banach algebra of power series and that* k *and* m *are non-negative integers with* $m > k$. *If* $S_{-m}(B)$ *is an algebra (a radical algebra) then* $S_{-k}(B)$ *is an algebra (a radical algebra).*

Proof. Since $S_{-i}(S_{-j}(B)) = S_{-(i+j)}(B)$, for all i and j, there is no loss in generality in assuming that $m = k + 1$. Suppose $S_{-(k+1)}(B)$ is an algebra. Then $S_{-k}(B)$ is the ideal $z(S_{-(k+1)})^{\#}$ in $S_{-(k+1)}$. Therefore S_{-k} is an algebra, and is radical if $S_{-(k+1)}$ is radical [19, Th. 1, p. 10]. This completes the proof.

If B is a Banach algebra of power series in which z is quasi-nilpotent, it is clear that B is a radical algebra, and therefore that $S_{-j}(B)$ is radical whenever it is an algebra [38, Th. (2.3.5), p. 57]. The next lemma shows that this is still nearly true for the generalized Banach algebras of power series we study.

LEMMA (3.6). *Suppose that* B *is a generalized Banach algebra of power series, that* z *is quasi-nilpotent in* B, *and that there is an integer* n *for which* fz^n *is a limit of polynomials for all* f *in* B. *Then:*

(A) *If* $S_{-1}(B)$ *is an algebra, then* B *is a radical algebra.*

(B) *If* m *and* k *are non-negative integers with* $m > k$, *and if* S_{-m} *is an algebra, then* S_{-k} *is a radical algebra.*

Proof. Part (B) will follow from applying Part (A) to $S_{-k}(B)$, in place of B; so we only prove Part (A).

Let R be the closure of the polynomials in B. Then R is a radical algebra [38, Th. (2.3.5), p. 57]. Suppose f is an arbitrary power series in B, and let $f = zg$. Then g belongs to the algebra $S_{-1}^{\#}$, so $f^{n+1} = z^n(zg^n)$ belongs to Bz^n, and hence to R. Therefore f^{n+1} and f are quasi-nilpotent, and the proof is complete.

The importance of assuming that some $S_{-j}(B)$ is an algebra or a radical algebra is that this assumption allows us to derive useful formulas involving products, quotients, powers, and roots of power series. There are a variety of minor variants of these formulas (see [9] [13] [14] [15]). In particular, the results below are quite similar to results in [14, Section 3]. The central idea behind all the proofs is the same; one writes $f = z^j g$, and works with g instead of f. We include the following lemmas in this paper mainly to familiarize the reader with the general techniques. Also, we will usually need to use the specific formulas in the following lemmas, instead of the variants from [13] [14] [15]. Lemma (3.7), in particular, will be used repeatedly in this paper. The results in [13] [14] [15] were used largely to get information about the ideal structure of $B^{\#}$. In particular we showed that the closed ideals in $B^{\#}$, when B is a Banach algebra of power series for which all $S_{-j}(B)$ are radical algebras, are the ideals $B^{(n)}$. More detailed information about the ideal structure of algebras of power series can be found in [14, Section 3] and [15, Section 4].

LEMMA (3.7). *Suppose that* B *is a generalized Banach algebra of power series and that* j *is a fixed positive integer. If* $S_{-j}(B)$ *is an algebra, then* fg *belongs to* B *whenever* f *belongs to* $(S_{-j})^{\#}$ *and* g *belongs to* $B^{(j)}$. *Moreover, there is a positive number* M, *independent of* f *and* g, *for which*

$$\|fg\| \leq M\|f\| \; \|g\|;$$

where the norms of fg *and* g *are taken in* B *and the norm of* f *is taken in* $(S_{-j})^{\#}$.

Proof. It will be enough to show that fg always belongs to B. For, in this case, the map from $(S_{-j})^{\#} \times B^{(j)}$ to B which takes the pair (f,g) to fg is separately continuous, and hence is jointly continuous [4, Prob. 4, p. 71] by the uniform boundedness principle.

Suppose therefore that f belongs to $(S_{-j})^{\#}$ and that $g = z^j h$ belongs to $B^{(j)}$. Since both f and h belong to the algebra $S_{-j}{}^{\#}$, fh belongs to $S_{-j}{}^{\#}$ as well. Hence $fg = (fh)z^j$ belongs to B, and the proof is complete.

LEMMA (3.8). *Suppose that* B *is a generalized Banach algebra of power series and that* $S_{-j}(B)$ *is a radical algebra. If* g *is a series in* B *of order* j, *and if* f *is an arbitrary series, then:*

fg *belongs to* $B \Longleftrightarrow f$ *belongs to* $(S_{-j})^{\#}$.

Proof. We already know, from Lemma (3.7), that fg belongs to B if f belongs to $(S_{-j})^{\#}$. Suppose, therefore, that fg belongs to B. Let $g = z^j h$. Then h and fh belong to $S_{-j}^{\#}$, and h has non-zero constant term. Since S_{-j} is radical, h^{-1} belongs to $(S_{-j})^{\#}$. Therefore $f = (fh)h^{-1}$ belongs to $(S_{-j})^{\#}$, and the proof is complete.

The final two lemmas of this section will not be needed until Section 9.

LEMMA (3.9). *Suppose that* B *is a generalized Banach algebra of power series and that* S_{-jn} *is an algebra. Then:*

(A) $(S^{(j)}_{-j(n-1)})^m \subseteq B$, *for all* $m \geq n$.

(B) $(S^{(j)}_{-j(n-1)})^n = B^{(jn)}$.

Proof. Suppose $f_1, f_2, \ldots f_m$ belong to $S^{(j)}_{-j(n-1)}$. Let $f_i = z^j f_i'$ for $i = 1,2,\ldots m$. Then $f_1 f_2 \ldots f_m = z^{jm}(f_1' f_2' \ldots f_m')$ belongs to B. Conversely, suppose f belongs to $B^{(jn)}$, and let $f = (z^j)^{n-1}h$. Since h belongs to $S^{(j)}_{-j(n-1)}$, this completes the proof.

LEMMA (3.10). *Suppose that* B *is a generalized Banach algebra of power series and that* f *is a series of order* jn *in* B. *If* S_{-jn} *is a radical Banach algebra, then the n'th roots of* f *all belong to* $S_{-j(n-1)}$.

Proof. Without loss of generality we may assume that $\pi_{jn}(f) = 1$, so that $f = z^{jn}(1+g)$. Since g belongs to the radical Banach algebra S_{-jn}, the n'th roots of $1 + g$, calculated by the binomial formula, all belong to $(S_{-jn})^{\#}$. Hence the n'th roots of f belong to $S_{-j(n-1)}$.

4. DERIVATIONS

In this section we discuss the derivations of generalized Banach algebras of power series. In our main result, Theorem (4.1), we prove that if $S_{-k}(B)$ is a radical algebra, and if gD is a derivation for some g of order k in B, then gD is a derivation for all g in $B^{(k)}$. When z^2D is a derivation this yields, in Corollary (4.3), the complete characterization of derivations stated in Theorem (2.6)(A). We also determine the spectra of derivations of B; and we prove several properties of derivations of generalized Banach algebras of power series which are analogous to properties proved by Kamowitz and Scheinberg [21] for $L^1(0,1)$. At the end of this section, in Lemma (4.8), we compare various hypotheses on derivations with conditions involving the sequence $\{\|z^n\|\}$.

THEOREM (4.1). *Suppose that* B *is a generalized Banach algebra of power series. If* $S_{-k}(B)$ *is a radical algebra, then the following are equivalent:*

(A) z^kD *is a derivation of* $B^{\#}$.

(B) gD *is a derivation of* $B^{\#}$ *for all* g *in* $B^{(k)}$.

(C) There is a series g *of order* k *in* B *for which* gD *is a derivation of* $B^{\#}$.

Proof. Suppose that z^kD is a derivation of $B^{\#}$ and that $g = z^kh$ belongs to $B^{(k)}$. Since h belongs to $(S_{-k})^{\#}$, $gDf = h(z^kDf)$ belongs to B for all f in $B^{\#}$; because of Lemma (3.7).

Conversely suppose g has order k and that gD is a derivation. By Lemma (3.8), Df belongs to $(S_{-k})^{\#}$ for all f in $B^{\#}$. Hence z^kD is a derivation of $B^{\#}$, and the proof is complete.

COROLLARY (4.2). *Suppose that* B *is a generalized Banach algebra of power series and that* $S_{-1}(B)$ *is a radical algebra. If* gD *is a derivation of* $B^{\#}$, *then* g *belongs to* $B^{(2)}$.

Proof. If g does not belong to $B^{(2)}$, then zD is a derivation of $B^{\#}$, by Theorem (4.1). Then zD is a bounded operator, by the closed graph theorem. But $zDz^n = nz^n$, so zD cannot be bounded, and the proof is complete.

The following complete characterization of derivations is now an immediate consequence of Theorem (4.1) and Corollary (4.2).

COROLLARY (4.3). *Suppose that* B *is a generalized Banach algebra of power series, that* $S_{-2}(B)$ *is a radical algebra, and that* z^2D *is a derivation of* $B^{\#}$. *Then* gD *is a derivation of* $B^{\#}$ *if and only if* g *belongs to* $B^{(2)}$.

The characterization of the derivations of $B^{\#}$ in Corollary (4.3), above, is roughly analogous to the characterization of derivations of $L^1(0,1)$ by Kamowitz and Scheinberg [21, Th. 2, p. 416]. The derivation z^2D on $B^{\#}$ plays the role of the derivation d on $L^1(0,1)$ defined by $(df)(x) = xf(x)$ [21, p. 417]. In fact, for the algebras we study in Section 13, which are both subalgebras of $L^1(0,1)$ and also algebraically homeomorphic to Banach algebras of power series, z^2D is just a restriction of the operator d. Though we will frequently point out analogies between Banach algebras of power series and $L^1(0,1)$, the correspondence is far from perfect. For instance, the spectrum of d is $[0,1]$ (see [21, Lemma, p. 422]) while its restriction to spaces of analytic functions commonly has the entire closed unit disc as its spectrum. Moreover, there are several cases where the structure of $L^1(0,1)$ does not seem to be known in sufficient detail to determine what analogies exist with Banach algebras of power series. In particular, is there a formula for the automorphisms of $L^1(0,1)$ which corresponds to the formula T_g in Banach algebras of power series?

The next theorem compares $B^{\#}$ and its space of derivations.

THEOREM (4.4). *Suppose that* B *is a generalized Banach algebra of power series and that the set of derivations of* B *is precisely the set of* gD *for* g *in* $B^{(k)}$. *Then the map* $g \to gD$ *is a linear homeomorphism from* $B^{(k)}$ *onto the set of derivations of* $B^{\#}$.

Proof. The map $gD \to g = gDz$ is a continuous linear bijection from the space of derivations onto $B^{(k)}$. Theorem (4.4) now follows from the open mapping theorem.

We will frequently need to know that certain derivations of $B^{\#}$ are compact. The following analogue of Theorem (4.1) will be useful for this purpose.

LEMMA (4.5). *Suppose that* B *is a generalized Banach algebra of power series. If* $S_{-k}(B)$ *is a radical algebra, then the following are equivalent:*

(A) z^kD *is a compact derivation of* $B^{\#}$.

(B) gD *is a compact derivation of* $B^{\#}$ *for all* g *in* $B^{(k)}$.

(C) There is a series g *of order* k *in* B *for which* gD *is compact.*

Proof. Suppose that z^kD is a compact operator on $B^{\#}$, and that $g = z^kh$ belongs to $B^{(k)}$. Then $gD = L_h(z^kD)$. L_h is a bounded operator on $B^{(k)} \supseteq z^kD(B^{\#})$, by Lemma (3.7), so gD is compact.

Conversely suppose g has order k and that gD is compact on $B^{\#}$. Since h has non-zero constant and S_{-k} is radical, h^{-1} belongs to $(S_{-k})^{\#}$. Hence $z^kD = L_{h^{-1}}(gD)$ is compact; and the proof is complete.

We will see, in Lemma (6.9) below, that, under the hypotheses we need to characterize the principal automorphisms of $B^{\#}$, L_z is always compact and hence z^3D is compact. Under the assumption that z^3D is compact, we can determine the spectra of the derivations of $B^{\#}$. The analogous result for $L^1(0,1)$ is [21, Lemma 11, p. 422].

THEOREM (4.6). *Suppose that* B *is a generalized Banach algebra of power series and that* $S_{-3}(B)$ *is a radical algebra. Suppose also that* z^2D *is a derivation of* $B^{\#}$ *and that* z^3D *is compact. If* g *belongs to* $B^{(2)}$ *and* $\pi_2(g) = \lambda$, *then* $sp(gD) = \lambda sp(z^2D)$.

Proof. Let $g = \lambda z^2 + h$. Then, by Lemma (4.5), $gD - \lambda z^2D = hD$ is compact. Since neither gD nor λz^2D can have any eigenvalues, it follows from the Fredholm theory that gD and z^2D have the same spectra. This completes the proof.

Of course the above theorem only reduces the problem of determining the spectrum of an arbitrary derivation to the problem of finding $sp(z^2D)$. In many cases, this is not too difficult. If z^2D is compact, or has a compact power, it is quasi-nilpotent, by the Fredholm theory. For $K<1/(n-1)!>$ and the spaces of analytic functions considered in Section 14, the spectrum of z^2D is the closed unit disc. In all cases z^2D is a weighted shift with respect to the bi-orthogonal system (z^n, π_n); so that information about its spectrum can be found in [6] and [30].

The next result on commuting derivations is analogous to [21, Th. 5, p. 419] for $L^1(0,1)$.

THEOREM (4.7). *Suppose that* B *is a radical generalized Banach algebra of power series and that there is a non-negative integer* k *for which* fz^k *is a limit of polynomials in* B, *for all* f *in* $B^{\#}$. *Then two derivations of* $B^{\#}$ *commute if and only if they are linearly dependent.*

Proof. Let the derivations be fD and gD. By Theorem (2.7), a derivation is completely determined by its value at z. Hence fD and gD commute if and only if the Lie product derivation $(fD)(gD) - (gD)(fD)$ maps z to 0. Since the Lie product maps z to $fDg - gDf$, the proof is complete.

The next simple result relates hypotheses on derivations with those on $\{\|z^n\|\}$.

LEMMA (4.8). *Suppose that* B *is a generalized Banach algebra of power series with* $\|z^n\| = |c_n|$, *and that* k *is a positive integer.*

(A) *If* $z^{k+1}D$ *is a derivation of* $B^\#$, *then* $\{nc_{n+k}/c_n\}_n$ *is bounded. For* $B = K^p\langle c_n\rangle$ *the condition is sufficient.*

(B) *If* $z^{k+1}D$ *is compact on* $B^\#$, *then* $\lim_n (nc_{n+k}/c_k) = 0$. *For* $B = K^p\langle c_n\rangle$, *the condition is sufficient.*

(C) *If* $\Sigma_n n\|\pi_n\| |c_{n+k}|$ *converges, then* $z^{k+1}D$ *is a compact map from* $B^\#$ *into* $K\langle c_n\rangle$.

Proof. Since $z^{k+1}D$ is bounded whenever it is a derivation of $B^\#$ and since $z^{k+1}Dz^n = nz^{k+n}$, Part (A) is clear. If $z^{k+1}D$ is compact, then $\lim_n z^{k+1}D(z^n/c_n) = \lim_n (nc_{n+k}/c_n) = 0$ in the norm topology of $B^\#$, since this limit is 0 in the Fréchet topology. We omit the easy proofs of the other assertions of the lemma.

By the above lemma and Theorem (4.1), if all $S_{-j}(B)$ are radical algebras and if none of the sequences $\{nc_{n+k}/c_k\}_n$ is bounded, then no gD can be a derivation of $B^\#$. Scheinberg [40, Th. 4, p. 322] proves this for $B = K\langle c_n\rangle$ without any assumptions about S_{-j}. The next theorem extends his result to B in which $\{\|\pi_n\|\ \|z^n\|\}$ is bounded (which is always true, in particular, if $\{z^n\}$ is a Schauder basis [31, Th. 3, p. 29]).

THEOREM (4.9). *Suppose that* B *is a generalized Banach algebra of power series with* $\|z^n\| = c_n$, *and that* $\{\|\pi_n\| c_n\}$ *is bounded. If there exists a series* g *of order* k+1 *in* B *for which* gD *is a derivation of* $B^{\#}$, *then* $\{nc_{n+k}/c_n\}_n$ *is bounded.*

Proof. Without loss of generality, we can assume $\pi_{k+1}(g) = 1$. Then, for all n, we have $\pi_{k+n}(gD(z^n)) = n$. Hence

$$n \leq \|\pi_{k+n}\| \ \|gD\| c_n \ = \ \|gD\| (c_n/c_{n+k}) \|\pi_{n+k}\| c_{n+k}$$

This completes the proof.

5. ENDOMORPHISMS

In this section we characterize the endomorphisms of certain Banach algebras of power series. We concentrate on endomorphisms which are not automorphisms. The study of automorphisms, which is much more difficult, is deferred to the next two sections. The basic technique we use is really quite simple. If g is a power series with zero constant term, then the operator T_g on $C[[z]]$ is given by the formula

$$(5.1) \qquad T_g = \sum g^n \otimes \pi_n;$$

where we use the standard notation $g^n \otimes \pi_n(f) = \pi_n(f)g^n$ (see [38, p. 64]). The series in formula (5.1) converges to T_g pointwise in the Fréchet topology. To show that T_g is an endomorphism of some $B^{\#}$ containing g, it will be enough to show that the series converges absolutely in the uniform operator topology on $B^{\#}$. To do this we first obtain, in Lemma (5.2), bounds on the sequence $\{\|g^n\|\}$. These same bounds are used in the converse problem of finding necessary conditions on g for T_g to be an endomorphism.

LEMMA (5.2). *Suppose that* B *and* $S_{-k}(B)$ *are radical generalized Banach algebras of power series and that* g *is a series in* $B^{(k)}$. *If* $\pi_k(g) = \lambda$, *then*

$$(5.3) \qquad \lim(\|g^n\|/\|z^{kn}\|)^{1/n} = |\lambda|$$

Proof. Let $g = z^k h$. Then h belongs to $S_{-k}^{\#}$, and $\pi_o(h) = \lambda$.

Since S_{-k} is radical, we have $\lim \|h^n\|^{1/n} = |\lambda|$; where the norm of $\|h^n\|$ is taken in $S_{-k}^{\#}$. By Lemma (3.7), there is a positive number M for which

$$\|g^n\| \le M \|z^{kn}\| \|h^n\|$$

for all n. Taking n'th roots and limits yields

$$\lim(\|g^n\|/\|z^{kn}\|)^{1/n} \le |\lambda| \tag{5.4}$$

In the special case $\lambda = 0$, this proves formula (5.3). If $\lambda \neq 0$, h is invertible in $S_{-j}^{\#}$, and $\lim\|h^{-n}\|^{1/n} = 1/|\lambda|$. Since $z^k = gh^{-1}$, we can now once more use Lemma (3.7) and reverse the inequality in formula (5.4). This completes the proof.

Before applying Lemma (5.2) to the study of the endomorphisms of $B^{\#}$, we mention the following consequence of Lemma (5.2) together with the arguments of Miziolek, Müldner, and Rek [32, pp. 44-47].

THEOREM (5.5). *Suppose that* B *is a generalized Banach algebra of power series and that* $S_{-1}(B)$ *is a radical algebra, and define* $d_n = \sup\{\|f_1 f_2 \cdots f_n\| : \|f_i\| \le 1\}$. *Then* $\lim(d_n/\|z^n\|)^{1/n}$ *exists and is finite.*

Section 11 below will be devoted to obtaining a variety of estimates on norms of products, like those in Lemma (5.2) and Theorem (5.5). No such estimates hold for $L^1(0,1)$, since it has a bounded approximate identity [32, p. 43]. Thus it would appear that our results on endomorphisms of generalized Banach algebras of power series have no natural analogue for $L^1(0,1)$.

The rest of this section is devoted to applying Lemma (5.2) to the study of endomorphisms.

THEOREM (5.6). *Suppose that* B *is a radical generalized Banach algebra of power series with* $\|z^n\| = c_n$ *and that* $S_{-2}(B)$ *is an algebra. If there exists a positive integer* m *for which* $B \subseteq S_{-m}(K<c_n>)$, *then* T_g *is a compact endomorphism of* $B^{\#}$ *for all* g *in* $B^{(2)}$.

Proof. Suppose g is a fixed series in $B^{(2)}$. By Lemma (5.2), there is a constant r for which $\|g^n\|$ is less than $r^n c_{2n}$ for all n. The assumption that $B \subseteq S_{-m}(K)$ implies that $\{\|\pi_n\| \, c_{n+m}\}_n$ is bounded. Since c_n is the norm of z^n in the algebra B, the series $\Sigma\|g^n \otimes \pi_n\|$ is dominated by the series $\Sigma_n r^n c_{n-m}$. But $\lim_n (c_{n-m})^{1/n} = \lim_n (c_n)^{1/n} = 0$, so that $\Sigma_n r^n c_{n-m}$ converges. Therefore $\Sigma\, g^n \otimes \pi_n$ converges in norm to T_g. This completes the proof.

THEOREM (5.7). *Suppose that* B *and* $S_{-1}(B)$ *are radical generalized Banach algebras of power series, and that* $R = \limsup(\|\pi_n\| \, \|z^n\|)^{1/n}$ *is finite. If* g *is a series in* B *for which* $|R\, \pi_1(g)| < 1$, *then* T_g *is a compact endomorphism of* $B^{\#}$.

Proof. As in the proof of Theorem (5.4), it will be enough to show that the series $\Sigma\|g^n \otimes \pi_n\|$ converges. But

$$\|g^n \otimes \pi_n\|^{1/n} = (\|\pi_n\| \, \|z^n\|)^{1/n} (\|g^n\|/\|z^n\|)^{1/n}.$$

Hence $\lim\|g^n \otimes \pi_n\|^{1/n} < 1$, by Lemma (5.2), and the proof is complete.

In the special case that $\{\|\pi_n\| \, \|z^n\|\}$ is bounded (for instance if $\{z^n\}$ is a Schauder basis), the previous theorem shows that T_g is an endomorphism of $B^{\#}$ for all g in B with $|\pi_1(g)| < 1$. The next result is a converse of this.

THEOREM (5.8). *Suppose that* B *and* $S_{-1}(B)$ *are radical generalized Banach algebras of power series, and that* g *is a series in* B. *If* T_g *is an endomorphism of* $B^{\#}$, *then* $|\pi_1(g)| \leq 1$. *If* T_g *is an automorphism of* $B^{\#}$, *then* $|\pi_1(g)| = 1$.

Proof. Suppose T_g is an endomorphism. Since $T_g(z^n) = g^n$, the sequence $\{\|g^n\|/\|z^n\|\}$ must be bounded. By Lemma (5.2), this requires that $|\pi_1(g)| \leq 1$. If T_g is an automorphism, the sequence $\{\|z^n\|/\|g^n\|\}$ must also be bounded. Therefore, by Lemma (5.2) again, $|\pi_1(g)| \geq 1$; and the proof is complete.

In Theorems (5.6) and (5.7), we showed that the endomorphisms T_g were all compact. In the case that T_g is compact, it is fairly easy to use the Fredholm theory to find the spectrum of T_g.

THEOREM (5.9). *Suppose that* B *is a radical generalized Banach algebra of power series and that* T_g *is a compact endomorphism of* $B^{\#}$. *If* $\pi_1(g) = \lambda$, *then the spectrum of* T_g *is precisely the set* $\{0\} \cup \{\lambda^n\}_0^\infty$.

Proof. Suppose that f is a series in $B^{\#}$, and let k be the order of f. If r is a non-zero scalar, then

$$\pi_k[(T_g - rI)(f)] = \pi_k(f)\,(\lambda^k - r). \tag{5.10}$$

Since T_g is compact, r cannot belong to the spectrum of T_g unless r is an eigenvalue. By formula (5.10), this requires that $r = \lambda^n$ for some n.

Conversely if $r = \lambda^n$, then formula (5.8) shows that the range of T_g contains no series of order k. This completes the proof.

6. AUTOMORPHISMS

The main purpose of this section is to prove the following theorem.

THEOREM (6.1). *Suppose that* B *is a generalized Banach algebra of power series. If*

(i) $S_{-3}(B)$ *is an algebra,*

(ii) z^2Df *converges in* B *for all* f *in* $B^{\#}$;

then T_g *is an automorphism of* $B^{\#}$ *for all* g *in* $z + B^{(2)}$.

We also prove a similar result, Theorem (6.4), with slightly different hypotheses. In Theorems (6.15) and (6.17), at the end of this section, we replace the hypothesis on z^2D with a hypothesis on $z^{m+1}D$ for arbitrary positive integers m. Theorems (6.15) and (6.17) also generalize a result of Ginsberg and Newman [8, Th. 1, p. 234] on generators of radical algebras.

Theorems (6.1) and (6.4) are the major results in this paper. We use the information about the automorphisms of $B^{\#}$ to determine the generators of $B^{\#}$, at the end of this section; to find bounds of $\{\|f^n\|\}$, in Section 10; to determine operators similar to shifts, in Section 12; and to study general quasi-nilpotent operators, in Section 14. Moreover, the lemmas and formulas we need for the proofs of Theorems (6.1) and (6.4) will themselves be used repeatedly in the remaining sections of this paper.

There are two parts to our proof that T_g is an automorphism of $B^{\#}$ for all g in $z + B^{(2)}$. First we show that all such T_g map $B^{\#}$ *into* itself (i.e., that such T_g are all endomorphisms); and then we show that all such T_g map $B^{\#}$ *onto* itself. The proof that T_g is an endomorphism of $B^{\#}$ for all g in $z + B^{(2)}$ is based on the same idea as the proof, in Section 5 above, of the analogous fact for g in $B^{(2)}$. We represent

T_g as the sum in $C[[z]]$ of a series of operators; and then we show that this series converges absolutely in the uniform operator topology for $B^\#$. Unfortunately we can no longer use the simple representation $T_g = \Sigma g^n \otimes \pi_n$. For, if g belongs to $z + B^{(2)}$, then

$$\|g^n \otimes \pi_n\| = \|g^n\| \, \|\pi_n\| \geq |\pi_n(g^n)| = 1;$$

so this series cannot converge in the uniform topology.

Instead, we let $g = z + z^2h$. Then, by adapting the usual proof of Taylor's theorem for functions given by convergent power series [2, Prop. 2.2, p. 37], [24, Section 4, pp. 443-445], we obtain

$$f\circ g = f\circ(z+z^2h) = f + \Sigma_{k=1}^{\infty} h^k z^{2k} D^k f/k!;$$

where the series converges in the Fréchet topology on $C[[z]]$. The above formula can easily be translated into the following formula involving operators on $C[[z]]$.

$$T_g - I = \Sigma_{k=1}^{\infty} (L_h)^k (z^{2k} D^k)/k!; \tag{6.2}$$

where I is the identity on $C[[z]]$, and where the series of operators converges pointwise in the Fréchet topology. We prove, in Lemma (6.5), that T_g is an endomorphism of $B^\#$ by showing that the series in formula (6.2) is an absolutely converging series of bounded linear operators on $B^\#$.

The proof that T_g is actually an automorphism for all g in $z + B^{(2)}$ is given in three lemmas. In Lemmas (6.9) and (6.10), respectively, we prove that T_g is an automorphism when g belongs to $z + B^{(3)}$ or when g has positive radius of convergence. Finally, in Lemma (6.11), we show that an arbitrary T_g can always be factored into a product of the special T_g considered in Lemmas (6.9) and (6.10).

After the following definition, we state Theorem (6.4), which gives alternate hypotheses under which T_g is an automorphism for all T_g in $z + B^{(2)}$.

DEFINITION (6.3). For each non-negative integer j, we define the operator Q_j on $C[[z]]$, and on its subspaces, by

$$Q_j(\Sigma_o^\infty \lambda_n z^n) = \Sigma_j^\infty \lambda_n z^n.$$

THEOREM (6.4). *Suppose that* B *is a generalized Banach algebra of power series. If*

(i) $S_{-3}(B)$ *is an algebra,*

(ii) z^2D *is a derivation of* $B^{\#}$,

(iii) $\lim_j L_z Q_j = 0$;

then T_g *is an automorphism of* $B^{\#}$ *for all* g *in* $z + B^{(2)}$.

Theorems (6.1) and (6.4) are not really independent results. After proving Theorem (6.4), we will prove, in Lemma (6.12), that if B satisfies the hypotheses of Theorem (6.1) then it also satisfies the hypotheses of Theorem (6.4). Usually the hypotheses of Theorem (6.1) are easier to verify than those of Theorem (6.4). In fact, if $\{z^n\}$ is a Schauder basis of $B^{\#}$, then z^2Df must necessarily converge in $B^{\#}$ whenever z^2Df belongs to $B^{\#}$. For $K^P<c_n>$, it is quite easy to verify that L_zQ_j converges to zero whenever z^2D is a derivation (just use the explicit formula in Lemma (4.8)(A)).

There are, however, important examples, in particular among the spaces of analytic functions we consider in Section 13, which fail to satisfy the hypotheses of Theorem (6.1), but which do satisfy the hypotheses of Theorem (6.4).

We now begin the proof of Theorem (6.4), by showing that each T_g maps $B^{\#}$ *into* itself.

LEMMA (6.5). *Suppose that* B *is a generalized Banach algebra of power series which satisfies the hypotheses of Theorem (6.4). Then:*

(A) *For all positive integers* k, $z^{2k}D^k$ *is a bounded linear operator from* $B^{\#}$ *to* $B^{(2)}$. *Moreover* $\lim(\|z^{2k}D^k\|/k!)^{1/k} = 0$.

(B) *If* g *belongs to* $z + B^{(2)}$, *then the series in formula (6.2) is an absolutely converging series of bounded linear operators on* $B^{\#}$; *and hence* T_g *is an endomorphism of* $B^{\#}$.

Proof. Let us first observe that (6.5)(B) will follow from (6.5)(A). Let $g = z + z^2h$; then L_h is a bounded linear operator on $B^{(2)}$, by Lemma (3.7). Thus if (6.5)(A) is true, the series in formula (6.2) does converge absolutely to the bounded linear operator $T_g - I$; and hence (6.5)(B) holds.

We now prove that $z^{2k}D^k$ maps $B^{\#}$ into itself. By hypothesis, z^2D maps $B^{\#}$ into itself. Hence, for each fixed j, z^2D maps $S_{-j}^{\#}$ into itself, because

$$(6.6) \qquad (z^2Df)z^j = z^2D(fz^j) - jz^{j+1}f.$$

Therefore, for each j, D maps $S_{-j}^{\#}$ into $S_{-(j+2)}^{\#}$. Hence D^k maps $B^{\#}$ into $S_{-2k}^{\#}$, and $z^{2k}D^k$ maps $B^{\#}$ into $B^{(2k)}$.

Let $\|\cdot\|_j$ be the norm in $S_{-j}^{\#}$, and also the operator norm on $S_{-j}^{\#}$. Recall, from Definition (2.4), that $\|f\|_j = \|fz^j\|$; so that L_{z^j} is an isometry from $S_{-j}^{\#}$ onto $B^{(j)}$. Thus we can use formula (6.6) to obtain:

$$(6.7) \qquad \|z^2D\|_j \le \|z^2D\| + j\|L_z\|_j \le \|z^2D\| + j\|L_zQ_j\|.$$

Let $M_j = \|z^2D\|_j$, which is the same as the norm of D as an operator from $S_{-j}^{\#}$ to $S_{-(j+2)}^{\#}$, since L_{z^2} is an isometry from $S_{-(j+2)}^{\#}$ onto $S_{-j}^{(2)}$. Similarly $\|z^{2k}D^k\|$ is the same as the norm of D^k as an operator from $B^{\#}$ to $S_{-2k}^{\#}$. Hence

$$\|z^{2k}D^k\| \le M_oM_1\ldots M_{2k-2}$$

and therefore (using [1, Prob. 12-4, p. 383]) we obtain

$$\limsup\,(\|z^{2k}D^k\|/k!)^{1/k} \le \limsup\,(M_{2k}/(k+1)).$$

But $\lim\,(M_{2k}/(k+1)) = 0$, by formula (6.7) together with the assumption that $\lim L_zQ_j = 0$. This completes the proof of Lemma (6.5).

We remark that the above proof does not use the full strength of hypothesis (6.4)(i) which says that $S_{-3}(B)$ is an algebra. Instead we use only the fact that S_{-2} is an algebra (which is a weaker assumption by Lemma (3.5)).

Before completing the proof of Theorem (6.4), we prove the important special case in which the hypothesis is strengthened by assuming that z^2D is compact. (Lemma (4.8) above gives conditions implying the compactness of z^2D, and also necessary conditions.) Not only is this special case much easier to prove, but also the conclusion, that $T_g - I$ is compact and quasi-nilpotent, is somewhat stronger. Theorem (6.8) will be important in Section 8.

THEOREM (6.8). *Suppose that* B *is a generalized Banach algebra of power series satisfying:*

(i) $S_{-2}(B)$ *is an algebra;*

(ii) z^2D *is a compact operator on* $B^{\#}$*;*

(iii) $\lim_j L_z Q_j = 0$.

Then $T_g - I$ *is compact and quasi-nilpotent for all* g *in* $z + B^{(2)}$.

Proof. Since L_{z^2} is a linear isometry from $S_{-2}^{\#}$ onto $B^{(2)}$, D is a compact operator from $B^{\#}$ to $S_{-2}^{\#}$. Hence $z^{2k}D^k = L_{z^{2k}}D^k$ is a compact operator on B. Therefore each of the summands of the series for $T_g - I$, in formula (6.2), is compact. By Lemma (6.5)(B), this series converges in the uniform norm, so $T_g - I$ is compact. Since $T_g - I$ cannot have any non-zero eigenvalues, it follows from the Fredholm theory that $T_g - I$ is quasi-nilpotent. This completes the proof of Theorem (6.8).

We are now ready to state the results, Lemmas (6.9), (6.10), and (6.11), which together allow us to complete the proof of Theorem (6.4). After stating these lemmas, we will show how Theorem (6.4) follows from them; and then we will prove each of the lemmas in turn.

LEMMA (6.9). *Suppose that* B *satisfies the hypotheses of Theorem (6.4). Then:*

(A) L_h *is a compact operator on* $B^{(2)}$, *for all* h *in* S_{-2}.

(B) $T_g - I$ *is a compact quasi-nilpotent operator on* $B^{\#}$ *for all* g *in* $z + B^{(3)}$.

LEMMA (6.10). *Suppose that* B *satisfies the hypotheses of Theorem (6.4) and that* g *belongs to* $z + B^{(2)}$. *If* g *is a power series with non-zero radius of convergence, then* T_g *is an automorphism of* $B^{\#}$.

LEMMA (6.11). *Suppose that* B *satisfies the hypotheses of Theorem (6.4) and that* g *belongs to* $z + B^{(2)}$. *Suppose also that* g_1 *is a series in* $z + B^{(2)}$ *for which* T_{g_1} *is an automorphism of* $B^{\#}$. *If* $g - g_1$ *belongs to* $B^{(k)}$, *with* $k \geq 3$, *then there is a series* g_2 *in* $z + B^{(k)}$ *for which* $T_g = T_{g_1} T_{g_2}$.

Proof of Theorem (6.4). Suppose $g = z + \Sigma_{n=2}^{\infty} \lambda_n z^n$ belongs to $z + B^{(2)}$. Let g_1 be $z + \lambda_2 z^2$. Then T_{g_1} is an automorphism of $B^{\#}$, by Lemma (6.10). By Lemma (6.11), we can find a series g_2 in $z + B^{(3)}$ such that $T_g = T_{g_1} T_{g_2}$. T_{g_2} is an automorphism, by Lemma (6.9); and thus T_g is a composition of two automorphisms of $B^{\#}$. Thus we will have proved Theorem (6.4) as soon as we prove Lemmas (6.9), (6.10), and (6.11).

Proof of Lemma (6.9). We first notice that L_z is a compact operator on $B^{\#}$; since it is the norm limit of the finite rank operators $L_z(I-Q_j)$. Now suppose h is an arbitrary series in S_{-2}, and let $h = zf$. Then f belongs to $S_{-3}^{\#}$, and $L_h = L_f L_z$. Since L_z is a compact operator from $B^{(2)}$ to $B^{(3)}$ and L_f is a bounded operator on $B^{(3)}$, because of Lemma (3.7); we see that L_h is a compact operator on $B^{(2)}$. This proves (6.9)(A).

Now suppose g belongs to $z + B^{(3)}$ and let $g = z + z^2 h$. Since $g - z$ belongs to $B^{(3)}$, h belongs to S_{-2}, instead of just to $S_{-2}^{\#}$. Thus each of the summands, $(L_h)^k (z^{2k} D^k)/k!$, of the series of formula (6.2) is a compact operator on $B^{\#}$. So $T_g - I$ is compact, and hence quasi-nilpotent. This completes the proof of Lemma (6.9).

Proof of Lemma (6.10). Since L_z is compact and has no eigenvalues, z is quasi-nilpotent; so $B^{\#}$ contains all series with positive radius of convergence. Let g' be the compositional inverse of g, that is, the series for which $g \circ g' = g' \circ g = z$. Then $T_g T_{g'}$ and $T_{g'} T_g$ are both the identity operator on $C[[z]]$. Since g' has a positive radius of convergence [2, prop. 9.1, p. 26], it must belong to $z + B^{(2)}$. By Lemma (6.5), both T_g and $T_{g'}$ map $B^{\#}$ into itself. Hence T_g and $T_{g'}$ are inverse automorphisms of $B^{\#}$, and Lemma (6.10) is proved.

Proof of Lemma (6.11). Since T_{g_1} is an automorphism of B, there is a series g_2 in B for which $g = T_{g_1}(g_2) = g_2 \circ g_1$. Thus we have $T_g = T_{g_1} T_{g_2}$. Moreover, $g_2 - z = T_{g_1}^{-1}(g - g_1)$ must belong to $B^{(k)}$, because T_{g_1} preserves the order of power series in $C[[z]]$. This completes the proof of Lemma (6.11), and therefore of Theorem (6.4) as well.

The only place in the proof of Theorem (6.4) that we use the assumption that $S_{-3}(B)$ is an algebra instead of the weaker assumption (according to Lemma (3.5)) that S_{-2} is an algebra is in the proof, in Lemma (6.9)(B), that L_h is compact when L_z is compact. If B is actually a Banach algebra of power series, then the polynomials are dense in B and in $S_{-2}(B)$. Hence, in this case, if h belongs to S_{-2}, then L_h is a compact operator on $B^{(2)}$, by Lemma (3.7); provided only that S_{-2} is an algebra. In a similar manner, if the polynomials are dense, we could prove that fD is a compact derivation of $B^{\#}$ for all f in $B^{(3)}$, provided that $z^2 D$ is a derivation, that L_z is compact, and that S_{-2} is an algebra (without assuming that S_{-3} is an algebra as we did in Lemma (4.5) and Theorem (4.6)). In fact, all those results in this paper which combine assumptions on $z^2 D$ with the assumption that $S_{-3}(B)$ is an algebra or a radical algebra are true under the weaker assumption that $S_{-2}(B)$ is an algebra; provided that B is a Banach algebra of power series instead of just a generalized Banach algebra of power series.

Notice also that we have not assumed that B is a radical algebra in the hypotheses of any of the theorems or lemmas in this section. We can conclude from the compactness of L_h for h in B, which is a special case of Lemma (6.9)(A), that B is in fact a radical algebra.

We now complete the proof of Theorem (6.1) by showing that any algebra which satisfies the hypotheses of Theorem (6.1) also satisfies the hypotheses of Theorem (6.4).

LEMMA (6.12). *Suppose that* B *is a generalized Banach algebra of power series. If* z^2Df *converges in* B *for all* f *in* $B^{\#}$, *then the sequence* $\{j\|L_zQ_j\|\}$ *is bounded.*

Proof. We need to consider the space *bv* of all complex sequences of bounded variation. This is the space of sequences $\{a_n\}_o^\infty$ for which

$$(6.13) \qquad |a_o| + \Sigma_{n=o}^{\infty} |a_{n+1} - a_n|$$

converges; where the norm of $\{a_n\}$ in *bv* is the sum of the series (6.13) (compare [4, p. 239]). We define the bi-linear map G from the Cartesian product of *bv* and C[[z]] to C[[z]] by

$$(6.14) \qquad G(\{a_n\},f) = \Sigma_{n=1}^{\infty} na_n \pi_n(f) z^{n+1}.$$

By our hypotheses the series $\Sigma_{n=1}^{\infty} n\pi_n(f) z^{k+1}$ converges in $B^{\#}$ for all f in $B^{\#}$. We can use the classical proof of Abel's convergence test for series of scalars [1, Th. 12-29, p. 365], without essential change, to prove that $G(\{a_n\},f)$ belongs to $B^{\#}$ for all $\{a_n\}$ in bv and all f in $B^{\#}$ (compare [43, Th. 5.2, p. 40]). G is then a bilinear map from $bv \times B^{\#}$ to $B^{\#}$. G is separately continuous by the closed graph theorem, and hence is jointly continuous by the uniform boundedness principle (see [4, Prob. 4, p. 71]).

Now fix j and define the sequence $\{b_n\}$ by

$$b_n = 0, \quad \text{if } n < j$$

$$b_n = 1/n, \quad \text{if } n \geq j.$$

Then $\{b_n\}$ has norm $1/j$ in bv, so that if f is an arbitrary series in $B^{\#}$, then

$$\|L_z Q_j f\| = \|G(\{b_n\},f)\| \leq \|G\|(1/j)\|f\|.$$

Therefore $\|L_z Q_j\| \leq \|G\|/j$. This completes the proof of Lemma (6.12), and also of Theorem (6.1).

The conclusion of the above lemma is significantly stronger than hypothesis (iii) of Theorem (6.4), which only requires that $\{\|L_z Q_j\|\}$ converge to zero. The stronger statement, that $\{j\|L_z Q_j\|\}$ is bounded, will be needed for some of the results in Section 10.

The previous results in this section have all involved assumptions on z^2D as an operator on $B^{\#}$. The mere fact that z^2D maps $B^{\#}$ into itself requires that the sequence $\{n\|z^{n+1}\|/\|z^n\|\}$ be bounded (see Lemma (4.8)(A)). This restriction is too severe for some applications. We now generalize Theorem (6.8) by replacing the hypotheses on z^2D and L_z with conditions on $z^{m+1}D$ and L_{z^m}, for an arbitrary positive integer m.

THEOREM (6.15). *Suppose that* B *is a generalized Banach algebra of power series and that* m *is a positive integer. If*

(*i*) $S_{-(m+1)}(B)$ *is an algebra,*

(*ii*) $z^{m+1}D$ *is a compact operator on* $B^{\#}$,

(*iii*) $\lim_j L_{z^m} Q_j = 0$;

then

(*A*) *For all* g *in* $z + B^{(m+1)}$, T_g *is an automorphism of* $B^{\#}$, *and* $T_g - I$ *is compact and quasi-nilpotent.*

(*B*) *If* B *is a Banach algebra of power series, it is topologically generated by each of its series of order one.*

Proof. Suppose $g = z + z^{m+1}h$ belongs to B. Then we have the following "Taylor expansion" analogous to formula (6.2):

$$T_g - I = \Sigma_{k=1}^{\infty} (L_h)^k (z^{(m+1)k} D^k)/k! \ , \tag{6.16}$$

where the series converges pointwise in the Fréchet topology. The suitable generalizations of formulas (6.6) and (6.7) are easy to prove; so the proof of Lemma (6.5) can be modified to show that the series in formula (6.16) converges in norm. The compactness of the summands in the series is proved as in Theorem (6.8). Hence $T_g - I$ is compact and quasi-nilpotent; and Part (A) is proved.

Suppose now that B is a Banach algebra of power series and that f is a series of order one in B. There is obviously a series g in $z + B^{(m+1)}$ which is a polynomial in f. By Part (A), T_g is an automorphism of B. Hence B is topologically generated by $g = T_g z$. Since g is in the algebra generated by f, the proof of Theorem (6.15) is complete.

It is not difficult to generalize our proofs of Theorems (6.1) and (6.4) to theorems involving $z^{m+1}D$ in place of z^2D. We omit the details and instead prove the following easy special case, which is adequate for many of our applications.

THEOREM (6.17). *Suppose that* B *is a Banach algebra of power series and that all* $S_{-j}(B)$ *are algebras. If there is a positive integer* k *for which* $z^{k+1}Df$ *converges in* $B^{\#}$ *for all* f *in* $B^{\#}$, *then* B *is generated by each of its series of order one.*

Proof. We first prove that $\lim_j L_{z^k} Q_j = 0$. This is done in essentially the same way that $\lim_j L_z Q_j = 0$ was proved in Theorem (6.12). It is only necessary to replace the bilinear map G, of formula (6.14), with the map G' defined by

$$G'(\{a_n\},f) = \Sigma_{n=1}^{\infty} na_n \pi_n(f)z^{n+k}.$$

Since $\lim_j L_{z^k} Q_j = 0$, L_{z^k} is a limit of finite rank operators and is therefore compact. Hence $z^{2k+1}D$ is compact; and Theorem (6.17) now follows from Theorem (6.15)(B), with $m = 2k$.

Using very different methods, Ginsberg and Newman [8] prove Theorem (6.17) in the special case that $B = K<c_n>$ when $\{c_n\}$ satisfies

(6.18) $\{c_{n+1}/c_n\}$ decreases monotonically

and

(6.19) $0 < c_{n+1}/c_n < 1/n^{\varepsilon}$, for some ε.

These results can be obtained from Theorem (6.17), or from Theorem (6.15); since formula (6.18) implies that all $S_{-j}(K<c_n>)$ are algebras [13, Th. (2.10), p. 645] and formula (6.19) implies that some $z^{k+1}D$ maps $K<c_n>$ into itself (by Lemma (4.8)(A)).

7. THE GROUP OF AUTOMORPHISMS

In this section, for the algebras of power series considered in the previous section, we study the group of automorphisms, considered as a topological group in the uniform operator topology. Recall, from Definition (2.3), that an automorphism T_g of $B^{\#}$ is said to be principal if g belongs to $z + B^{(2)}$, and is said to be scalar if $g = \lambda z$ for some scalar λ. In Theorem (7.1) we show that the group of principal automorphisms is topologically anti-isomorphic to $z + B^{(2)}$, which is a group under power series composition. In Theorem (7.2), we show that the group of scalar automorphisms is discrete in the uniform operator topology. And, in Theorem (7.3), we show that the group of automorphisms is a topological semi-direct product of the subgroup of principal automorphisms with the subgroup of scalar automorphisms. The hypotheses of Theorems (7.1) and (7.3) are identical to the hypothesis of Theorem (6.4). Because of Lemma (6.12), these theorems would also hold for algebras satisfying the hypothesis of Theorem (6.1).

At the end of this section we discuss which T_h are automorphisms of $B^{\#}$; but we will only be able to give a complete answer when $\{z^n\}$ is an unconditional basis.

THEOREM (7.1). *Suppose that* B *is a generalized Banach algebra of power series. If*

(i) $S_{-3}(B)$ *is an algebra,*

(ii) z^2D *is a derivation of* $B^{\#}$,

(iii) $\lim_j L_z Q_j = 0$;

then

(A) $z + B^{(2)}$ *is a topological group under the operation of formal power series composition.*

(B) *The map* $g \to T_g$ *is a homeomorphic group anti-isomorphism from* $z + B^{(2)}$ *onto the group of principal automorphisms of* $B^{\#}$.

Proof. By Theorem (6.4), T_g is an automorphism for all g in $z + B^{(2)}$, so it is clear that $z + B^{(2)}$ is a group and that the map $g \to T_g$ is a group anti-isomorphism. Moreover, the inverse map $T_g \to g = T_g z$ is continuous from the uniform operator topology on the group of principal automorphisms to the norm topology on $z + B^{(2)}$. Thus to complete the proof of the theorem, we need only show that the map $g \to T_g$ is continuous.

We first prove continuity at z, the group identity in $z + B^{(2)}$. Suppose g belongs to $z + B^{(2)}$, and let $g = z + z^2 h$. Then, by Lemma (6.5),

$$\|T_g - I\| \le \Sigma_{k=1}^{\infty}(\|z^{2k}D^k\|/k!)\|L_h\|^k,$$

where L_h is considered as an operator on $B^{(2)}$. Because of (6.5)(A), the continuity of $g \to T_g$ at $g = z$ will follow if we can show that $\|L_h\|$ becomes small when g is near z. By Lemma (3.7), there is a number M for which $\|L_h\|$ is always less than M times the norm of h in $S_{-2}^{\#}$. But the norm of h in $S_{-2}^{\#}$ is just $\|g-z\|$ (see Definition (2.4)). Hence $\|L_h\| \le M\|g-z\|$, and the continuity at $g = z$ is proved.

Now suppose that f is an arbitrary fixed series in $z + B^{(2)}$. Given $\varepsilon > 0$, we must show that $\|T_g - T_f\| < \varepsilon$ when g is close to f in $z + B^{(2)}$. Let f' be the compositional inverse of f, and choose $\delta > 0$ such that $\|T_h - I\| < \varepsilon/\|T_f\|$ whenever $\|h-z\| < \delta\|T_{f'}\|$. Suppose now that $\|g-f\| < \delta$, then

$$\|g \circ f' - z\| = \|T_{f'}(g-f)\| < \delta\|T_{f'}\|.$$

Therefore

$$\|T_g - T_f\| = \|T_f(T_{gof'} - I)\| < \varepsilon.$$

This completes the proof of Theorem (7.1).

THEOREM (7.2). *If* B *is a generalized Banach algebra of power series, then the group of scalar automorphisms of* B *is discrete in the uniform operator topology.*

Proof. Suppose that $T_{\lambda z}$ is a scalar automorphism with $\lambda \neq 1$. Then

$$\|T_{\lambda z}-I\| \geq \sup_n \|(T_{\lambda z}-I)z^n\|/\|z^n\| = \sup_n |\lambda^n-1| \geq \sqrt{2}.$$

This completes the proof of Theorem (7.2).

It is not hard to show that $\sqrt{2}$ in the above proof can be replaced by $\sqrt{3}$; and in fact, if λ is not a root of unity, $\|T_{\lambda z}-I\| \geq 2$ (see [22, p. 269]). One obvious consequence of Theorem (7.1) is that the map $\lambda \to T_{\lambda z}$ cannot be continuous if there are infinitely many scalar automorphisms. We are now ready to discuss the full automorphism group as a topological group in the uniform operator topology.

THEOREM (7.3). *Suppose that* B *is a generalized Banach algebra of power series which satisfies the hypothesis of Theorem (7.1). Then:*

(A) The automorphism group of $B^{\#}$ *is, algebraically and topologically, a semi-direct product of the subgroup of principal automorphisms with the subgroup of scalar automorphisms.*

(B) The principal automorphism subgroup is the component of the identity in the full automorphism group.

Proof. The principal automorphism subgroup is connected, by Theorem (7.1); and the scalar automorphism subgroup is discrete, by Theorem (7.2). Part (B) will therefore follow from Part (A); so we prove (A).

It is clear that the intersection of the principal automorphism subgroup with the scalar automorphism subgroup is the identity automorphism. Also it is easy to show that every automorphism is a product of a principal automorphism and a scalar automorphism. In fact, suppose h belongs to B and $\pi_1(h) = \lambda \neq 0$; let $h = \lambda g$; then

$$T_h = T_{\lambda g} = T_g T_{\lambda z}. \tag{7.4}$$

Moreover, if λ is a scalar and g belongs to $z + B^{(2)}$, then $T_{(z/\lambda)} T_g T_{\lambda z}(z)$ belongs to $z + B^{(2)}$. So the principal automorphism subgroup is normal, and hence the full automorphism subgroup is an algebraic semi-direct product of the principal automorphism group and the scalar automorphism group [18, p. 6].

To complete the proof of the theorem, we must show that the automorphism group has the product topology. To do this, it will be enough to show that the projection map $T_{\lambda g} \to T_g$ is continuous [18, pp. 58-59] [33, pp. 96-98], where g belongs to $z + B^{(2)}$ and $T_{\lambda g}$ is an automorphism of $B^{\#}$. The map $T_{\lambda g} \to T_{\lambda g} z = \lambda g$ is clearly continuous. Moreover, since $T_{\lambda g}$ is an automorphism, $|\lambda| = 1$, by Theorem (5.8), and hence the map $\lambda g \to (\lambda g)/\pi_1(\lambda g) = g$ is also continuous. Finally, the map $g \to T_g$ is continuous by Theorem (7.1). This completes the proof of Theorem (7.3).

Notice that the algebraic part of the above theorem involves no hypotheses on B other than that it be a generalized Banach algebra of power series. The following obvious consequence of formula (7.4) is often useful.

COROLLARY (7.5). *Suppose that* B *is a radical generalized Banach algebra of power series for which* T_g *is an automorphism of* $B^{\#}$ *for all* g *in* $z + B^{(2)}$. *If* h *belongs to* B *and* $\pi_1(h) = \lambda \neq 0$; *then* T_h *is an automorphism (endomorphism) of* $B^{\#}$ *if and only if* $T_{\lambda z}$ *is an automorphism (endomorphism) of* $B^{\#}$.

The above corollary reduces the problem of determining the automorphisms of $B^{\#}$ to the problem of determining when $T_{\lambda z}$ is an automorphism of $B^{\#}$. The question of when $T_{\lambda z}$ is an endomorphism or automorphism of $B^{\#}$ is completely independent of the algebra structure of $B^{\#}$; so, unfortunately, the techniques we have developed in this paper are of little use in answering this question. Instead, we must rely on general results from the theory of bi-orthogonal sequences. For any series f, $T_{\lambda z}f = \Sigma\lambda^n \pi_n(f)z^n$. So, $T_{\lambda z}$ is an endomorphism of $B^{\#}$ if and only if $\{\lambda^n\}$ is a multiplier [43, Def. 5.1, p. 40] of the bi-orthogonal sequence (z^n,π_n); and $T_{\lambda z}$ is an automorphism if and only if both $\{\lambda^n\}$ and $\{\lambda^{-n}\}$ are multipliers of (z^n,π_n). The scalar automorphisms are easy to determine when $\{z^n\}$ is an unconditional basis.

THEOREM (7.6). *Suppose that* B *is a radical Banach algebra of power series and that* T_g *is an automorphism of* $B^{\#}$ *for all* g *in* $z + B^{(2)}$. *Suppose also that* $\{z^n\}$ *is an unconditional basis for* B. *Then* T_h *is an automorphism of* $B^{\#}$ *if and only if* h *is in* B *and* $|\pi_1(h)| = 1$.

Proof. Since $\{z^n\}$ is an unconditional basis, every bounded sequence of scalars is a multiplier of (z^n,π_n), by [43, Th. 16.5, p. 484]. Therefore Theorem (7.6) follows from Corollary (7.5) and Theorem (5.8).

Since every bounded sequence is a multiplier of (z^n,π_n) in $K^{\infty}\langle c_n\rangle$, the above proof is also valid for $B = K^{\infty}<c_n>$.

In the special case that λ is a primitive k'th root of unity, it is clear that $T_{\lambda z}$ is an automorphism of $B^{\#}$ whenever it is an endomorphism. Moreover, in this case the question of whether $T_{\lambda z}$ is an automorphism has a convenient reformulation in terms of the structure of $B^{\#}$. For each $j = 0,1,2,\ldots,k-1$, define P_j as the natural projection from $C[[z]]$ onto the set of those series f for which $\pi_n(f) = 0$ unless $n \equiv j \pmod{k}$. Then an easy calculation shows that $T_{\lambda z}$ is an automorphism of $B^{\#}$ if and only if all P_j map $B^{\#}$ into itself.

The final result in this section generalizes Scheinberg's necessary condition for $K<c_n>$ to have a non-scalar automorphism [40, Th. 5, p. 322] to more general algebras. Our proof is modeled on his.

THEOREM (7.7). *Suppose that* B *is a radical Banach algebra of power series with unconditional basis* $\{z^n\}$ *and that* T_h *is an automorphism of* $B^{\#}$. *If* $h - \pi_1(h)z$ *has order* k, *then* $\{n\|z^{n+k-1}\|/\|z^n\|\}$ *is bounded.*

Proof. Let $\pi_1(h) = \lambda$ and $h = \lambda g$. Since $\{z^n\}$ is an unconditional basis, $T_{\lambda z}$ is an automorphism of $B^{\#}$; hence $T_g = T_h T_{\lambda z}^{-1}$ is also an automorphism. Since $g - z$ has order k, we have

$$\pi_{n+k-1} T_g(z^n) = n\, \pi_k(g).$$

The proof is completed by observing that $\pi_k(g)$ is a constant, T_g is a bounded operator, and $\{\|\pi_n\|\ \|z^n\|\}$ is bounded.

Notice that if $\{\|z^n\|\ \|\pi_n\|\}$ is bounded, our proof of Theorem (7.7) is still valid when T_h is a principal automorphism, even if $\{z^n\}$ is not an unconditional basis.

8. EXPONENTIALS OF DERIVATIONS

If V is a bounded linear operator on a Banach space, then the exponential of V is the bounded operator defined by

$$\exp V = \Sigma_{k=0}^{\infty} V^k/k! .$$

It is well known, and easy to show, that if V is a derivation of a Banach algebra, then $\exp V$ is an automorphism. In this section we investigate, for Banach algebras of power series, which automorphisms are exponentials of derivations or are products of exponentials of derivations.

If an automorphism T of a Banach algebra is the exponential of a derivation ∂, then it belongs to the continuous one-parameter group of automorphisms $\{\exp(t\partial)\}_t$ (see [48]). Thus the group of automorphisms generated by the exponentials of derivations is connected. In Theorem (8.5) below, we will see that, for Banach algebras of power series of the sort we have been studying, the group of principal automorphisms is precisely the group generated by the exponentials of derivations. For Banach algebras of power series, we can use the homeomorphic group anti-isomorphism $g \to T_g$ to construct continuous one-parameter subgroups of the group $z + B^{(2)}$ under composition. If $T_h = \exp(gD)$, and if λ is a scalar, we let $h^{(\lambda)} = \exp(\lambda gD)z$. Then $h^{(\lambda)} \circ h^{(\mu)} = h^{(\lambda+\mu)}$; and the map $\lambda \to h^{(\lambda)}$ is continuous (compare [39, p. 325]).

Kamowitz and Scheinberg [21] [22] [40] have used exponentials of derivations of commutative semi-simple Banach algebras to study the spectra of the automorphisms of these algebras. In Theorem (8.8) we apply our results on exponentials of derivations to determine the spectra of principal automorphisms of Banach algebras of power series. Our results are quite similar to results proved for $L^1(0,1)$ by Kamowitz and Scheinberg [21].

The following lemma collects a few of the known facts about exponentials of derivations on arbitrary Banach algebras. For additional related results about derivations and automorphisms of general Banach algebras see [21] and [48]. For related results about the Fréchet algebra $C[[z]]$, see [39].

LEMMA (8.1). *If* B *is a Banach algebra, then the map* $V \to \exp V$ *maps the quasi-nilpotent derivations of* B *homeomorphically onto the set of automorphisms* T *for which* T-I *is quasi-nilpotent.*

Proof. If T is an automorphism for which T-I is quasi-nilpotent, then $\log T$, defined by the usual power series, is a derivation [21, Th. 6, p. 420]. The rest of the lemma can be proved by simple arguments on power series, or can be considered as a special case of standard results about the operational calculus [4, pp. 568-571 and 586].

For algebras satisfying the hypothesis of Theorem (6.8), we can combine Lemma (8.1) with results from Sections 4 and 6 to completely characterize exponentials of derivations. Recall, from Definition (6.3), that Q_j is the natural projection from $B^{\#}$ onto $B^{(j)}$; and recall, from Lemma (6.12), that $\lim_j L_z Q_j = 0$ if $z^2 Df$ converges in $B^{\#}$ for all f in $B^{\#}$.

THEOREM (8.2). *Suppose that* B *is a generalized Banach algebra of power series satisfying:*

(i) $S_{-2}(B)$ *is an algebra;*

(ii) $z^2 D$ *is a compact operator on* $B^{\#}$;

(iii) $\lim_j L_z Q_j = 0$.

Then the map $gD \to \exp(gD)$ *is a homeomorphism from the space of derivations of* $B^{\#}$ *onto the group of principal automorphisms.*

Proof. All derivations are compact, and hence quasi-nilpotent, by Corollary (4.3) and Lemma (4.5). If T_g is a principal automorphism, then $T_g - I$ is quasi-nilpotent, by Theorem (6.8). Theorem (8.2) now follows from Lemma (8.1).

Zeller-Meier [48] shows that if T is a bounded automorphism of a Banach algebra, and if the spectrum of T is contained in the open right half-plane, then T is the exponential of a derivation. The following theorem is a sharper result for principal automorphisms of generalized Banach algebras of power series. A rough analogue for $L^1(0,1)$ of the following theorem is [21, Th. 14, p. 423].

THEOREM (8.3). *Suppose that* B *is a generalized Banach algebra of power series. If* T_h *is a principal automorphism of* $B^{\#}$ *whose spectrum does not disconnect the plane, then there is a derivation* gD *on* $B^{\#}$ *for which* $\exp(gD) = T_h$.

Proof. Using the usual contour-integral-based operational calculus [4, pp. 568-570], we can find a logarithm V of T_h. This V is a bounded linear operator on $B^{\#}$ for which $\exp V = T_h$ [4, Th. 11, p. 570].

Construct a sequence $\{d_n\}$ of positive numbers for which: d_{n+1}/d_n decreases monotonically; $\Sigma n d_{n+1}/d_n$ converges; and $\Sigma \|\pi_n\| d_n$ converges. Then $B \subseteq K\langle d_n\rangle$ and, by [13, Th. (2.10), p. 645], $K\langle d_n\rangle$ satisfies the hypothesis of Theorem (8.2). Hence $gD = \log(T_h)$ is a derivation on $K\langle d_n\rangle^{\#}$. Since B is continuously embedded in $K\langle d_n\rangle$, V is just the restriction of gD from $K\langle d_n\rangle^{\#}$ to $B^{\#}$. This completes the proof.

Using Theorem (6.15) in place of Theorem (6.8), we obtain the following generalization of Theorem (8.2).

THEOREM (8.4). *Suppose that* B *is a generalized Banach algebra of power series and that* m *is a positive integer. If*

(i) $S_{-(m+1)}(B)$ *is an algebra,*

(ii) $z^{m+1}D$ *is a compact operator on* $B^{\#}$,

(iii) $\lim_j L_{z^m}Q_j = 0$.

Then the map exp *maps the space of derivations* gD *with* g *in* $B^{(m+1)}$ *homeomorphically onto the group of* T_h *with* h *in* $z + B^{(m+1)}$.

Proof. Suppose g is any formal power series and $\exp(gD) = T_h$. It is easy to see that g and h-z have the same order [39, Prop. 3, p. 324]. Except for the use of Theorem (6.15) in place of Theorem (6.8), the rest of the proof of Theorem (8.4) is identical to the proof of Theorem (8.2).

The hypothesis on B in the following theorem is identical with the hypotheses of Theorems (6.1), (7.1), and (7.3). The analogue for $L^1(0,1)$ of the following theorem is [21, Th. 9 and Lemma 13, pp. 422-423].

THEOREM (8.5). *Suppose that* B *is a generalized Banach algebra of power series satisfying:*

(i) $S_{-3}(B)$ *is an algebra.*

(ii) z^2D *is a derivation of* $B^{\#}$.

(iii) $\lim_j L_zQ_j = 0$.

If h *is a series in* $z + B^{(2)}$ *with* $\pi_2(h) = \lambda$, *then there is a unique series* g *in* $B^{(3)}$ *for which*

$$T_h = [\exp(\lambda z^2D)]\,[\exp(gD)]. \tag{8.6}$$

Proof. A simple calculation shows that $\exp(\lambda z^2D) = T_{h_1}$ where

$$h_1 = z + \lambda z^2 + \lambda^2 z^3 + \dots = z/(1-\lambda z).$$

Since $h-h_1$ belongs to $B^{(3)}$, there is, by Lemma (6.11), a series h_2 in $z + B^{(3)}$ for which $T_h = T_{h_1} T_{h_2}$. By Lemma (6.9), $T_h - I$ is quasi-nilpotent; so, by Lemma (8.1), there is a derivation gD with $\exp(gD) = T_{h_2}$. It is easy to see that g belongs to $B^{(3)}$ [39, Prop. 3, p. 324]. Since formula (8.6) clearly determines gD uniquely, the proof is complete.

Combining Theorem (8.5) with our earlier characterization of principal automorphisms, in Theorem (7.1), we obtain:

COROLLARY (8.7). *Suppose that* B *is a generalized Banach algebra of power series satisfying the conditions in Theorem (8.5). If* T *is an automorphism of* $B^{\#}$, *then the following are equivalent:*

(A) Tz *belongs to* $z + B^{(2)}$.

(B) T *belongs to the component of the identity of the automorphism group of* $B^{\#}$.

(C) T *belongs to the group of automorphisms generated by the exponentials of derivations of* $B^{\#}$.

(D) T *is a product of two automorphisms, each of which is the exponential of a derivation of* $B^{\#}$.

We now calculate the spectra of principal automorphisms. The analogue for $L^1(0,1)$ is [21, Lemma 11, p. 422]. Recall again that the hypotheses of Theorems (6.4), (7.1), (7.3), and (8.5) are identical, and are weaker than the hypothesis of Theorem (6.1).

THEOREM (8.8). *Suppose that* B *is a generalized Banach algebra of power series satisfying the hypotheses of Theorem (8.5), and suppose that* h *is a series in* $z + B^{(2)}$ *with* $\pi_2(h) = \lambda$. *Then*

$$sp(T_h) = \exp[\lambda\ sp(z^2D)].$$

Proof. Using formula (8.6), we have

$$T_h - \exp(\lambda z^2 D) = [\exp(\lambda z^2 D)]\ [\exp(gD) - I];$$

so $T_h - \exp(\lambda z^2 D)$ is compact, by Lemma (6.9). The only eigenvalue of T_h or of $\exp(\lambda z^2 D)$ is 1. Hence, it follows from the Fredholm theory for compact operators that T_h and $\exp(\lambda z^2 D)$ have the same spectra. This completes the proof of Theorem (8.8).

We now obtain some information about the spectra of automorphisms which are not principal.

THEOREM (8.9). *Suppose that* B *is a radical generalized Banach algebra of power series. If* h *is a series in* $z + B^{(2)}$ *for which* $T_h - I$ *is compact, and if* λ *is a scalar of absolute value* 1 *for which* $T_{\lambda z}$ *is an automorphism of* $B^{\#}$, *then* $sp(T_{\lambda h}) = sp(T_{\lambda z})$.

Proof. $T_{\lambda h} - T_{\lambda z}$ is the compact operator $(T_h - I)T_{\lambda z}$. Also, it is clear that the set of eigenvalues of $T_{\lambda z}$ is the set $\{\lambda^k\}_o^\infty$. Let $g = \lambda h$. Using formula (5.10), as in the proof of Theorem (5.9), we see that T_g can have no eigenvalues which are not integral powers of λ. Hence for $r \neq \lambda^k$, we have, by the Fredholm theory for compact operators, that r belongs to the spectrum of $T_{\lambda h}$ if and only if r belongs to the spectrum of $T_{\lambda z}$. Finally, by formula (5.10) again, we see that, for all non-negative integers k, $T_g - \lambda^k I$ can have no series of order k in its range. Hence all λ^k belong to the spectrum of T_g, and the proof is complete.

Even in the favorable case that $T_h - I$ is compact, the above theorem merely reduces the problem of determining the spectrum of $T_{\lambda h}$ to that of determining the spectrum of $T_{\lambda z}$, for $|\lambda| = 1$. If λ is a primitive k'th root of unity, it is not difficult to show that the spectrum of $T_{\lambda z}$ is precisely the set $\{1,\lambda,\ldots\lambda^{k-1}\}$. If λ is not a primitive root of unity, the spectrum of $T_{\lambda z}$ contains the unit circle $\{r: |r| = 1\}$, since this is the closure of the set of eigenvalues of $T_{\lambda z}$. For general $|\lambda| = 1$, r belongs to the resolvent of $T_{\lambda z}$ if and only if the sequence $\{(\lambda^n - r)^{-1}\}$ is a multiplier of the bi-orthogonal sequence (z^n, π_n). Thus, if $\{z^n\}$ is an unconditional basis of B, and if λ is not a root of unity, then it follows from [43, Th. 16.5, p. 484] that the spectrum of $T_{\lambda z}$ is precisely the unit circle.

9. DERIVATIONS AND AUTOMORPHISMS OF IDEALS

In this section we obtain formulas for the derivations, automorphisms, and endomorphisms of the ideals $B^{(j)}$ in a generalized Banach algebra of power series B. Recall that, if B is a radical Banach algebra of power series and all $S_{-j}(B)$ are algebras, then $\{B^{(j)}\}$ is the complete collection of closed non-zero ideals of $B^{\#}$ [14, Lemma (3.2), p. 173] [15, Th. (4.1)].

The arguments in this section are in themselves fairly simple, but they rely heavily on the results of the previous sections. Our determination of the derivations, endomorphisms and automorphisms of $B^{(j)}$ proceeds in several steps. In Theorem (9.1), we show that derivations and endomorphisms of $B^{(j)}$ must all be of the form gD and T_g, respectively, for suitable g. In Theorem (9.4) we identify the derivations and automorphisms of $B^{(j)}$ with those of $S_{-(j-1)}$; and, in Lemma (9.5), we prove a similar but weaker result for endomorphisms. In Lemma (9.6) we show that conditions on z^2D, $L_{z^m}Q_j$, $K<c_n>$, etc. in the hypotheses of the various theorems in the previous sections about derivations, endomorphisms, and automorphisms of $B^{\#}$, imply the corresponding hypotheses for all $S_{-k}(B)$. The actual theorems on derivations, automorphisms, and endomorphisms of $B^{(j)}$ (Theorems (9.7), (9.8), (9.9) and (9.10)) are then easy consequences of the corresponding theorems for $S_{-(j-1)}^{\#}$.

Just as for $B^{\#}$, the deepest and most useful result is the characterization of automorphisms. Under suitable hypotheses on $B^{\#}$, we show, in Theorem (9.7), that if g belongs to $z^j + B^{(j+1)}$, then there is an automorphism T of $B^{(j)}$ for which $Tz^j = g$. In Section 10 this result will be applied to obtain estimates of norms of both powers and products of elements of B. In Section 12, we apply our knowledge of the automorphisms of $B^{(j)}$ to determine operators which are similar to restrictions of a given weighted shift.

The hypothesis of the following theorem is essentially the same as that of Theorem (2.7). The hypothesis of the theorem is clearly satisfied by all Banach algebras of power series; and it is also satisfied by any generalized Banach algebra of power series on which some $z^k D$ is a derivation (see Lemma (3.1)).

THEOREM (9.1). *Suppose that* B *is a generalized Banach algebra of power series and that* j *is a positive integer. If there is an integer* k *for which* fz^k *is a limit of polynomials in* B *whenever* f *belongs to* $B^{\#}$*; then:*

(A) *If* ∂ *is a derivation of* $B^{(j)}$*, then* $\partial = gD$ *for some* g *in* $S_{-(j-1)}$*;*

(B) *If* V *is an endomorphism of* $B^{(j)}$*, then* $V = T_g$ *for some* g *with* g^j *in* $B^{(j)}$.

Proof. We first prove (A). Since ∂z^j belongs to $B^{(j)}$, it is a power series of order at least j. Hence there is a series g, with zero constant term, such that

$$\partial(z^j) = jz^{j-1}g = gD(z^j). \tag{9.2}$$

Since $\partial(z^j)$ belongs to B, formula (9.2) implies that g belongs to $S_{-(j-1)}$. Just as in the proof of Theorem (2.7), we will have $\partial = gD$ if we prove that $\partial(z^k) = gD(z^k)$ for all $k \geq j$. From formula (9.2) we obtain

$$\partial(z^{jk}) = \partial((z^j)^k) = jk\, z^{j(k-1)} z^{j-1} g.$$

Since ∂ is a derivation, we also have

$$\partial(z^{jk}) = \partial((z^k)^j) = j z^{k(j-1)} \partial(z^k).$$

Equating the two values of $\partial(z^{jk})$, we obtain:

$$\partial(z^k) = kz^{k-1} g = gD(z^k).$$

This completes the proof of Part (A).

We now prove (B). Since $[V(z^j)]^{j+1} = [V(z^{j+1})]^j$, the order of the power series $V(z^{j+1})$ is greater than that of the series $V(z^j)$. Hence there is a series g, with zero constant term, for which

$$V(z^{j+1}) = g\, V(z^j). \tag{9.3}$$

If we take the j'th power of both sides of formula (9.3) and then divide by $V(z^{j^2})$, we obtain:

$$V(z^j) = g^j \quad \text{and} \quad V(z^{j+1}) = g^{j+1}.$$

As above, it will be enough to show that $V(z^k) = T_g(z^k) = g^k$ for all $k \geq j$. Now

$$V(z^{(j+1)k}) = g^{(j+1)k}, \qquad \text{and}$$

$$V(z^{(j+1)k}) = V(z^{jk})V(z^k) = g^j V(z^k).$$

Equating the two values of $V(z^{(j+1)k})$, we obtain $V(z^k) = g^k = T_g(z^k)$, for all $k \geq j$. This completes the proof of Theorem (9.1).

Notice that the derivation gD is completely determined by $gD(z^j)$ alone; but that the endomorphism T_g is determined by $T_g(z^j)$ and $T_g(z^{j+1})$ together.

We can now identify the derivations and automorphisms of $B^{(j)}$ with those of $S_{-(j-1)}$.

THEOREM (9.4). *Suppose that* B *is a generalized Banach algebra of power series and that* $S_{-j}(B)$ *is a radical algebra. Then*

(A) gD *is a derivation of* $B^{(j)}$ *if and only if it is a derivation of* $S_{-(j-1)}$.

(B) Every automorphism (endomorphism) of $S_{-(j-1)}$ *is an automorphism (endomorphism) of* $B^{(j)}$. *If* g *has order* 1, *then* T_g *is an automorphism (endomorphism) of* $B^{(j)}$ *if and only if it is an automorphism (endomorphism) of* $S_{-(j-1)}$.

Proof. By Lemma (3.9), $[S_{-(j-1)}]^j = B^{(j)}$, so any derivation, endomorphism, or automorphism of $S_{-(j-1)}$ must also be one of $B^{(j)}$.

Conversely, suppose that gD is a derivation of $B^{(j)}$. From Theorem (9.1) we know that g belongs to $S_{-(j-1)}$. Suppose that f is an arbitrary series in $S_{-(j-1)}$ and let $f = zh$. Then

$$(gDf)z^{j-1} = gD(fz^{j-1}) - h(j-1)gz^{j-1}.$$

$gD(fz^{j-1})$ belongs to B, since gD is a derivation of $B^{(j)}$; and

$h(j-1)gz^{j-1}$ belongs to B, by Lemma (3.7). This completes the proof of Part (A).

Now suppose that g is a series of order one and that T_g is an endomorphism of $B^{(j)}$. Then g belongs to $S_{-(j-1)}$, by Lemma (3.10); and zg^{j-1} is a series of order j in B, by Lemma (3.9). Now suppose f is an arbitrary series in $S_{-(j-1)}$, and let $T_g(f) = zh$. Then $T_g(fz^{j-1}) = h(zg^{j-1})$ belongs to $B^{(j)}$, since T_g is an endomorphism of $B^{(j)}$. Hence h belongs to $S_{-j}^{\#}$, by Lemma (3.8), so $T_g(f)$ belongs to $S_{-(j-1)}$. Thus T_g is an endomorphism of $S_{-(j-1)}$.

Now suppose that T_g is an automorphism and that g' is the compositional inverse of g. Then $T_g^{-1} = T_{g'}$ is an endomorphism of $B^{(j)}$, and hence, by the above argument, $T_{g'}$ is also an endomorphism of $B_{-(j-1)}$. This completes the proof of Theorem (9.4).

For endomorphisms T_g for which g has order greater than 1, we have only the following weaker result.

LEMMA (9.5). *Suppose that* B *is a generalized Banach algebra of power series, that* g *is a series of order* $k \geq 1$, *and that* $S_{-jk}(B)$ *is a radical algebra. Then:*

(A) If T_g *is an endomorphism of* $S_{-k(j-1)}$, *then* T_g *is an endomorphism of* $B^{(j)}$.

(B) If T_g *is an endomorphism of* $B^{(k)}$, *then* g *belongs to* $S_{-k(j-1)}$.

Proof. Suppose that T_g is an endomorphism of $S_{-k(j-1)}$. By Lemma (3.9), $[S^{(k)}_{-k(j-1)}]^j = B^{(jk)}$; so T_g is an endomorphism of $B^{(jk)}$. Moreover, if $m \geq j$, then $T_g(z^m) = g^m$ belongs to B, by Lemma (3.9) again. Hence T_g is an endomorphism of $B^{(j)}$.

Conversely, suppose that T_g is an endomorphism of $B^{(j)}$. Then $T_g(z^j) = g^j$ belongs to B. Hence g belongs to $S_{-k(j-1)}$, by Lemma (3.10). This completes the proof of Lemma (9.5).

Theorem (9.4) reduces the problem of determining the derivations and automorphisms of the ideal $B^{(j)}$ to that of finding the derivations and automorphisms of the generalized Banach algebra of power series $S_{-(j-1)}$. We already know, from Sections 4 and 6, conditions on $S_{-(j-1)}$ under which we can determine all its derivations and principal automorphisms. The following lemma shows that these conditions on $S_{-(j-1)}$ are automatic consequences of the corresponding conditions on B.

LEMMA (9.6). *Suppose that* B *is a generalized Banach algebra of power series and that* j *and* m *are non-negative integers. Let* c_n *be the norm of* z^n *in* B, *and let* d_n *be the norm of* z^n *in* $S_{-j}(B)$. *Then:*

(A) $S_{-m}(S_{-j}(B)) = S_{-(m+j)}(B)$.

(B) z^2D *is a derivation of* B *if and only if it is a derivation of* $S_{-j}(B)$.

(C) $B \subseteq S_{-m}(K<c_n>)$ *if and only if* $S_{-j}(B) \subseteq S_{-m}(K<d_n>)$.

(D) $\{L_{z^m}Q_k\}$ *converges to zero as a sequence of operators on* $B^{\#}$ *if and only if it converges to zero as a sequence of operators on* $S_{-j}(B)^{\#}$.

(E) If $\lim_k L_{z^m}Q_k = 0$, *then* $z^{m+1}D$ *is a compact operator on* $B^{\#}$ *if and only if it is a compact operator on* $S_{-j}(B)^{\#}$.

(F) $\{kL_zQ_k\}$ *is a bounded sequence of operators on* $B^{\#}$ *if and only if it is a bounded sequence of operators on* $S_{-j}(B)^{\#}$.

Proof. Parts (A) and (C) are immediate consequences of Definition (2.4); and Part (B) is an immediate consequence of formula (6.6). Parts (D) and (F) follow from the fact that L_{z^j} is an isometry from $S_{-j}^{\#}$ onto $B^{(j)}$. Since $\lim_k L_{z^m} Q_k = 0$ implies that L_{z^m} is compact, Part (E) follows the obvious analogue of formula (6.6) in which $z^{m+1}D$ replaces z^2D. This completes the proof of Lemma (9.6).

It is now easy to combine the results of this section with results in previous sections to characterize the automorphisms, derivations, and endomorphisms of $B^{(j)}$. The previous results in this section have only involved hypotheses on various $S_{-k}(B)$. To these hypotheses, we will need to add additional hypotheses; these additional hypotheses are the same ones we used in previous sections to characterize the automorphisms, derivations, and endomorphisms of $B^{\#}$. We start with the automorphisms since they are the most important.

THEOREM (9.7). *Suppose that* B *is a generalized Banach algebra of power series and that* j *is a positive integer. If*

(i) $S_{-(j+2)}(B)$ *is an algebra,*

(ii) z^2D *is a derivation of* $B^{\#}$,

(iii) $\lim_k L_z Q_k = 0$;

then

(A) *If* g *belongs to* $z + S_{-(j-1)}^{(2)}$, *then* T_g *is an automorphism of* $B^{(j)}$ *and of* $S_{-(j-1)}$.

(B) *If* h *belongs to* $z^j + B^{(j+1)}$, *then there is a unique automorphism* T_g *of* $B^{(j)}$ *for which* $T_g(z^j) = h$ *and* $\pi_1(g) = 1$.

Proof. By Lemma (9.6), we see that $S_{-(j-1)}(B)$ satisfies all the hypotheses of Theorem (6.4). Hence, by Theorem (6.4), T_g is an automorphism of $S_{-(j-1)}$ for all g in $z + S_{-(j-1)}^{(2)}$. Hence, by Theorem (9.4), T_g is an automorphism of $B^{(j)}$.

Suppose that h belongs to $z^j + B^{(j+1)}$. Then there is a unique g for which $g^j = h$ and $\pi_1(g) = 1$. This g belongs to $S_{-(j-1)}$, by Lemma (3.10). T_g is an automorphism of $B^{(j)}$, by Part (A); so Theorem (9.7) is proved.

THEOREM (9.8). *Suppose that* B *is a generalized Banach algebra of power series and that* j *is a positive integer. If* $S_{-(j+1)}(B)$ *is a radical algebra and if* z^2D *is a derivation of* $B^{\#}$, *then:*

(A) gD *is a derivation of* $B^{(j)}$ *if and only if* g *belongs to* $S^{(2)}_{-(j-1)}$.

(B) If h *belongs to* $B^{(j+1)}$, *there is a unique derivation* gD *of* $B^{(j)}$ *for which* $gD(z^j) = h$.

Proof. By Lemma (9.6), $S_{-(j-1)}(B)$ satisfies the hypotheses of Corollary (4.3). Part (A) then follows from Corollary (4.3) and Theorem (9.4). Suppose now that h belongs to $B^{(j+1)}$, and define g by $jz^{j-1}g = h$. Then g belongs to $S^{(2)}_{-(j-1)}$. This completes the proof.

THEOREM (9.9). *Suppose that* B *and* $S_{-j}(B)$ *are radical generalized Banach algebras of power series and that* $R = \limsup (\|\pi_n\| \, \|z^n\|)^{1/n}$ *is finite. Then:*

(A) T_g *is a compact endomorphism of* $B^{(j)}$ *for all* g *in* $S_{-(j-1)}$ *with* $|R\,\pi_1(g)| < 1$.

(B) If h *is a series in* $B^{(j)}$ *with* $0 < |R^j\,\pi_j(h)| < 1$, *then there are precisely* j *distinct automorphisms,* T_g, *of* $B^{(j)}$ *with* $T_g(z^j) = h$.

Proof. By Lemma (9.6), we see that $S_{-(j-1)}$ satisfies all the hypotheses of Theorem (5.7). Part (A) therefore follows from Theorems (5.7) and (9.4). If h is a series of order j in $B^{(j)}$, then there are j

distinct power series whose j'th power is h. These j'th roots all belong to $S_{-(j-1)}$, by Lemma (3.10). This completes the proof.

THEOREM (9.10). *Suppose that* B *and all* $S_{-n}(B)$ *are radical generalized Banach algebras of power series and that there is an integer* m *for which* $B \subseteq S_{-m}(K<\|z^n\|>)$. *Then:*

(A) *If* g *is a series of order* $k > 1$; *then* T_g *is an endomorphism of* $B^{(j)}$ *if and only if* g^j *belongs to* $B^{(j)}$. *Also,* T_g *is an endomorphism of* $B^{(j)}$ *if and only if it is an endomorphism of* $S_{-k(j-1)}(B)$.

(B) *If* h *is a series in* B *of order* jk, *with* $k > 1$, *then there are precisely* j *automorphisms of* $B^{(j)}$ *which map* z^j *to* h.

Proof. By Lemma (9.6), all $S_{-n}(B)$ satisfy the hypotheses of Theorem (5.6). Part (A) now follows from Lemmas (3.10) and (9.5) together with Theorem (5.6). Part (B) follows from Part (A) and Lemma (3.10). This completes the proof of Theorem (9.10).

Theorems (9.7), (9.8), (9.9), and (9.10) generalize some of our earlier results characterizing the derivations, endomorphisms, and automorphisms of $B^{\#}$ to similar results for arbitrary $B^{(j)}$ and $S_{-k}(B)$. Most of the other results we proved for $B^{\#}$ can similarly be generalized. For instance, suppose that B is a Banach algebra of power series with an unconditional basis and that B satisfies hypotheses (i) and (ii) of Theorem (9.7). Then we could generalize Theorem (7.6) to show that there is an automorphism of $B^{(j)}$ which maps z^j to h in $B^{(j)}$ if and only if $|\pi_j(h)| = 1$.

10. RATES OF QUASI-NILPOTENCE

Suppose that B is a generalized Banach algebra of power series, and that f is a power series for which f^n belongs to B for all sufficiently large n. In this section we obtain a variety of estimates of the rate of growth of the sequence $\{\|f^n\|\}$; and we also consider the analogous problem for sequences $\{\|f_1 f_2 \dots f_n\|\}$, where the f_i are chosen from some fixed finite set. The sharpness of our different estimates will depend on differing assumptions on B. We will always assume that some or all $S_{-j}(B)$ are radical algebras. Recall, from Lemma (3.9) and (3.10), that if all $S_{-j}(B)$ are algebras, then f belongs to some S_{-j} if and only if f^n belongs to B for all sufficiently large n.

Rates of growth of sequences of powers and products of elements in an arbitrary radical Banach algebra (or even in an arbitrary Fréchet algebra) are studied in [32], [44], and [47]. These papers investigate, in particular, the relationship between the rates of growth of sequences $\{\|x^n\|\}$ and $\{\|x_1 x_2 \dots x_n\|\}$ and the properties of functions defined on A by power series. The relationship between $\{x^n\}$ and $\{x_1 x_2 \dots x_n\}$ when the elements of A are algebraically nilpotent is discussed in [10].

In Lemma (5.2) above, we obtained estimates on the sequence $\{\|f^n\|\}$ for f in B (see also Theorem (5.5)). In fact, all our results about endomorphisms of $B^{\#}$ were based on these estimates on $\|f^n\|$. As one application of the sharper estimates we obtain in this section, we find, in Corollary (10.8), a lower bound on the distance between cosets of the principal automorphism subgroup for $B^{\#}$. Also, much of Section 14, which studies operators $f(T) = \Sigma\lambda_n T^n$, for arbitrary quasi-nilpotent T, is based on the results of the present section.

Our first, and weakest, estimate assumes only that sufficiently many $S_{-j}(B)$ are radical algebras. We normalize to the case that the leading coefficient of all the series is 1.

THEOREM (10.1). *Suppose that* B *is a generalized Banach algebra of power series for which* $S_{-j}(B)$ *is a radical algebra, and that* P *is a finite subset of* $z + S^{(2)}_{-(j-1)}$. *If*

$$d_n = \sup\{\|f_1 f_2 \ldots f_n\| : f_i \in P\}, \quad \textit{and}$$

$$b_n = \inf\{\|f_1 f_2 \ldots f_n\| : f_i \in P\};$$

then $$\lim(b_n/\|z^n\|)^{1/n} = \lim(d_n/\|z^n\|)^{1/n} = 1$$

Proof. First, recall from Lemma (3.9) that if $n \geq j$, then all products of n elements of P are in $B^{(j)}$. For each f in P, define f' in $1 + S_{-j}$ by $f = zf'$. Each $L_{f'}$ is a bounded operator on $B^{(j)}$, by Lemma (3.7); and, since S_{-j} is radical, $\lim\|L_{f'}\|^{1/n} = 1$, for all f in P. So given $\varepsilon > 0$, there exist an M such that

$$\|L_{f'_1} L_{f'_2} \ldots L_{f'_n}\| \leq M(1+\varepsilon)^n$$

for all $n \geq j$. If $f_1, f_2, \ldots f_n$ belong to P,

$$f_1 f_2 \ldots f_n = L_{f'_1} L_{f'_2} \ldots L_{f'_n} z^n .$$

Hence

$$d_n/\|z^n\| \leq M(1+\varepsilon)^n.$$

The lower bound for $b_n/\|z^n\|$ is obtained in a similar manner from the lower bound on the operator $L_{f'}$.

This completes the proof.

The requirement that the series in P have order 1 is just a normalization. If P contains a series f of order $k+1$, we can let $f = z^k h$ and form a new set P' with z and h. Instead of doing this in general, we give the following important special case.

COROLLARY (10.2). *Suppose that* B *is a generalized Banach algebra of power series and that* f *is a series in* $S_{-j}^{(k)}$. *If* $S_{-(j+k)}$ *is a radical algebra and if* $\pi_k(f) = \lambda$, *then*

$$\lim_n (\|f^n\|/\|z^{kn}\|)^{1/n} = |\lambda| .$$

Proof. Let $f = z^k g$. If $\lambda \neq 0$, Corollary (10.2) follows from applying Theorem (10.1) to the set $P = \{z,(g/\lambda)\}$. For $\lambda = 0$, L_g is quasinilpotent and $f^n = (L_g)^n z^{kn}$. This completes the proof.

We obtain estimates on $\|f^n\|$, instead of just on $\|f^n\|^{1/n}$, by applying our results on automorphisms from Section 9.

THEOREM (10.3). *Suppose that* B *is a generalized Banach algebra of power series and that* f *is a series of order* k *in* S_{-j}, *with* $\pi_k(f) = \lambda$. *If*

(i) $S_{-(j+k+2)}$ *is an algebra,*

(ii) $z^2 D$ *is a derivation of* $B^{\#}$,

(iii) $\lim_i L_z Q_i = 0$;

then the sequence $\{\|f^n\|/(|\lambda|^n \|z^{kn}\|)\}_n$ *is bounded above and below.*

Proof. Without loss of generality, we can assume $\lambda = 1$. Applying Lemma (3.10) to the radical generalized Banach algebra of power series S_{-j}, we can find a series g in $z + S_{-(j+k-1)}^{(2)}$ with $g^k = f$. By Theorem (9.7),

T_g is an automorphism of $B^{(j+k)}$. Moreover, $T_g(z^{kn}) = g^{kn} = f^n$. Since T_g is bounded and has a bounded inverse, this completes the proof.

We now wish to generalize the estimate on $\{\|f^n\|\}$, in Theorem (10.3), to estimates on $\{\|f_1 f_2 \cdots f_n\|\}$ which are sharper than the estimates in Theorem (10.1). To do this we will have to show that the automorphism T_g, for fixed g, has a single common upper bound on the sequence of Banach spaces $\{S_{-j}\}_o^\infty$; we will also need to prove the existence of universal bounds on derivations and on left multiplications. If B is a generalized Banach algebra of power series for which all S_{-j} are radical algebras, it follows from Theorem (9.4) that every automorphism and every derivation of $B^\#$ maps each $S_{-j}^{(k)}$ into itself. Similarly, if f belongs to $S_{-k}^\#$, then L_f maps each $S_{-j}^{(k)}$, $0 \le j < \infty$, into itself, by Lemma (3.7).

LEMMA (10.4). *Suppose that* B *is a generalized Banach algebra of power series and that* f *belongs to* $S_{-k}(B)$. *If all* $S_{-j}(B)$ *are radical algebras, then there is an* $M > 0$ *such that for all* j, L_f, *as an operator on* $S_{-j}^{(k)}$, *has norm less than* M.

Proof. It follows from the definition of the norm of S_{-j} (see Definition (2.4)) that the norm of L_f as an operator on $S_{-j}^{(k)}$ is the same as the norm of the restriction of L_f to $B^{(j+k)}$. This completes the proof.

To obtain a universal bound on derivations and automorphisms we will need to replace the assumption that $\lim_k L_z Q_k = 0$ with the stronger assumption that $\{kL_z Q_k\}_k$ is bounded. Recall from Lemma (6.12) that $\{kL_z Q_k\}$ is bounded if $z^2 Df$ converges in $B^\#$ for all f in $B^\#$.

LEMMA (10.5). *Suppose that* B *is a generalized Banach algebra of power series and that* V *is a derivation or a principal automorphism of* $B^\#$. *If*

(i) All $S_{-j}(B)$ *are algebras,*

(ii) z^2D *is a derivation of* $B^{\#}$,

(iii) $\{kL_zQ_k\}$ *is a bounded sequence of operators on* $B^{\#}$;

then there is an $M > 0$ *such that, for all* j, *the operator* V *acting on* $S^{\#}_{-j}$ *has norm less than* M.

Proof. z^2D has a common bound, as an operator on each $S^{\#}_{-j}$, by Formula (6.7). Since the most general derivation of $B^{\#}$ has the form $L_h z^2D$, for h in $S^{\#}_{-2}$, each derivation has a common bound on all $S^{\#}_{-j}$, by Lemma (10.3).

Suppose T_g is a principal automorphism of $B^{\#}$, and hence of all $S^{\#}_{-j}$. Let $g = z + z^2h$ and let N be a common bound for L_h and z^2D. Using Formula (6.2), and recalling that the norm of z^2D on $S^{\#}_{-j}$ is the same as the norm of D as an operator from $S^{\#}_{-j}$ to $S_{-(j+2)}^{\#}$, we obtain:

$$\|T_g\| \leq \Sigma_{k=o}^{\infty} \|L_h\|^k \|z^{2k}D^k\|/k! \leq e^{N^2}.$$

Thus, for all $j \geq 0$, e^{N^2} is a bound for T_g as an operator on $S^{\#}_{-j}$. This completes the proof.

Notice that we can combine Lemmas (9.6) and (10.5) to get bounds on T_g for g in some $z + S^{(2)}_{-m}$. Applying Lemma (10.5) to S_{-m} instead of B, we see that T_g will have a common upper bound as an operator on $S^{\#}_{-j}$ for all $j \geq m$. By Theorem (9.1), T_g is also an automorphism on the m additional spaces $B^{(m+1)}$, $S^{(m)}_{-1}, \ldots, S^{(2)}_{-(m-1)}$; and we can choose the bound for T_g acting on $\{S^{\#}_{-j}\}_{j\geq m}$ to also be a bound on these m additional spaces. Also notice that the universal upper bound for the inverse automorphism T_g^{-1} provides a universal lower bound for T_g. With these remarks to Lemma (10.5), we are now ready to obtain the promised estimate on $\{\|f_1f_2\ldots f_n\|\}$.

As in Theorem (10.1), we normalize to the case that all series have z as their leading term.

THEOREM (10.6). *Suppose that* B *is a generalized Banach algebra of power series and that* P *is a finite subset of* $z + S_{-m}^{(2)}$. *Let*

$$d_n = \sup\{\|f_1 f_2 \cdots f_n\| : f_i \in P\}, \quad \textit{and}$$

$$b_n = \inf\{\|f_1 f_2 \cdots f_n\| : f_i \in P\}.$$

If

(i) *All* $S_{-j}(B)$ *are algebras,*

(ii) $z^2 D$ *is a derivation of* $B^{\#}$,

(iii) $\{kL_z Q_k\}$ *is a bounded sequence of operators on* $B^{\#}$;

then the sequences $\{b_n/\|z^n\|\}$ *and* $\{d_n/\|z^n\|\}$ *are bounded above and are bounded away from zero.*

Proof. Throughout this proof we let $\|f\|_j$ be the norm of f in $S_{-j}^{\#}$, and we put $\|f\|_j = \infty$ if f does not belong to $S_{-j}^{\#}$. By Lemma (3.9) $\|f_1 f_2 \cdots f_n\|_j$ is finite if all f_i belong to S_{-m} and $n > m$. Theorem (10.6) is a special case of the following result which we will prove:

(*) There are positive numbers $N(P)$ and $M(P)$, which depend only on the set P, such that for all $j \geq 0$, we have:

$$N(P)\|z^n\|_j \leq \|f_1 f_2 \cdots f_n\|_j \leq M(P)\|z^n\|_j;$$

whenever all f_i belong to P.

We prove (*) by an induction on the number of elements in P. If P has only one element f, then $\dot{T}_f^{-1}(f^n) = z^n$, so (*) follows from the fact that T_f has universal upper and lower bounds. Suppose that (*) holds for all sets of k elements, and suppose that P has $k+1$ elements. We can replace P by $T_f^{-1}(P)$ for some P in f without changing the validity of (*). We can therefore assume that $P = \{z\} \cup P'$, where P' is a set of k series in $z + S_{-m}^{(2)}$. An arbitrary product of elements of P can be written as $f_1 f_2 \ldots f_n z^i$, where $f_1, f_2, \ldots, f_n$ belong to P'. For all $j \geq 0$, we have:

$$\|f_1 f_2 \ldots f_n z^i\|_j = \|f_1 f_2 \ldots f_n\|_{i+j} \leq M(P')\|z^n\|_{i+j} = M(P')\|z^{n+i}\|_j,$$

by induction; and similarly for the lower bound on the product. This completes the proof of (*), and hence of Theorem (10.6).

We can substantially improve the estimates in Theorems (10.3) and (10.6) if we add the assumption that $\lim(n\|z^{n+1}\|/\|z^n\|) = 0$. For $B = K^p\langle c_n\rangle$, this added assumption is equivalent to z^2D being a compact operator (see Lemma (4.8)). We give the analogue only of Theorem (10.3), since this is the result we will need in applications.

THEOREM (10.7). *Suppose that* B *is a generalized Banach algebra of power series and that* f *is a series of order* k *in* S_{-j}, *with* $\pi_k(f) = \lambda$. *If*

(i) *All* $S_{-m}(B)$ *are algebras,*

(ii) z^2D *is a derivation of* $B^{\#}$,

(iii) $\{nL_zQ_n\}$ *is a bounded sequence of operators on* $B^{\#}$,

(iv) $\lim n\|z^{n+1}\|/\|z^n\| = 0$;

then

(A) $\lim_n \|f^n - \lambda^n z^{kn}\|/\|\lambda^n z^{kn}\| = 0.$

(B) $\lim \|f^n\|/\|z^{kn}\| = |\lambda|$

Proof. Part (B) follows easily from Part (A), so we only need to prove (A). Without loss of generality we can assume $\lambda = 1$. Also, since we can replace f by its k'th root (see Lemma (3.10)), we can assume that f belongs to $z + S_{-j}^{(2)}$. Let $f = z + z^2h$; then

$$f^n - z^n = h(f^{n-1}z^2 + f^{n-2}z^3 + \ldots z^{n+1}).$$

L_h is a bounded operator on $B^{(j+2)}$, by Lemma (3.7); and $f^{n-i}z^{i+1}$ belongs to $B^{(j+2)}$ for all $n \geq j + 1$, by Lemma (3.9). So, by applying Theorem (10.6) to the set $\{f,z\}$, we see that there is a positive number M such that

$$\|f^n-z^n\| \leq nM\|z^{n+1}\| = \|z^n\| \; (nM\|z^{n+1}\|/\|z^n\|).$$

This completes the proof of Theorem (10.7).

In Theorem (7.3), we showed that, under suitable hypotheses, the cosets of the principal automorphism subgroup of the automorphism group of $B^{\#}$ are open and closed. For algebras satisfying the hypotheses of Theorem (10.7), we can find a lower bound for the distance between cosets.

COROLLARY (10.8). *Suppose that* B *is a generalized Banach algebra of power series which satisfies the hypotheses of Theorem (10.7) and that* T_g *and* T_h *are automorphisms of* $B^{\#}$. *If* $\pi_1(h)/\pi_1(g) = \lambda$, *then* $\|T_g-T_h\| \geq \limsup |\lambda^n-1|$.

Proof. For all n

$$\begin{aligned}\|T_g - T_h\| &\geq \|(T_g - T_h)z^n\|/\|z^n\| \\ &= \|g^n - h^n\|/\|z^n\| \\ &\geq |\lambda^n - 1|(\|g^n\|/\|z^n\|) - (\|\lambda^n g^n - h^n\|/\|z^n\|).\end{aligned}$$

But, by Theorem (10.7), $\lim\|g^n\|/\|z^n\| = 1$ and $\lim\|\lambda^n g^n - h^n\|/\|z^n\| = 0$. This completes the proof.

If λ is not a root of unity, or if λ is a root of -1, it is easy to see that $\limsup|\lambda^n - 1| = 2$. For all λ with $|\lambda| = 1$, $\limsup|\lambda^n - 1| \geq \sqrt{3}$.

11. ORBITS OF THE AUTOMORPHISM GROUP

Suppose that B is a generalized Banach algebra of power series and that g and h are series of order j for which $\pi_j(g) = \pi_j(h)$. In Theorem (9.7), we saw that, under suitable hypotheses on B, there is an automorphism of $B^{(j)}$ that maps g to h. In this section we investigate when there is a principal automorphism of all of $B^{\#}$ which maps g to h. Unfortunately our results are somewhat fragmentary. We will be able to give a complete answer in some special cases; but we can make only a plausible conjecture for the case of arbitrary g and h.

We will see that results about the orbits of the principal automorphism group of $B^{\#}$ can be used to find normal subgroups of $z + B^{(2)}$ (recall that $z + B^{(2)}$ is a group under formal power series composition if T_g is an automorphism of $B^{\#}$ for all g in $z + B^{(2)}$). These results should be useful in the study of conjugacy classes of formal power series in B (see [39] for a discussion of conjugacy classes in $C[[z]]$ and in the algebra of convergent power series).

Since algebra automorphisms are linear maps, we can, without loss of generality, consider only orbits of series with leading coefficient one. It will often be convenient to express a series whose leading term is z^j as the j'th power of a series whose leading term is z. Recall from Lemmas (3.9) and (3.10) that the series in $z^j + B^{(j+1)}$ coincide with the series which are of the form h^j for h in $z + S^{(2)}_{-(j-1)}$. If h is a series with leading term z, we will denote the *compositional inverse* of h by $\tilde{h}$; i.e., $\tilde{h}$ is defined by $h \circ \tilde{h} = \tilde{h} \circ h = z$, or by $T_{\tilde{h}} = T_h^{-1}$.

Our first theorem shows that the orbit of z^j under the principal automorphism group is easy to describe.

THEOREM (11.1). *Suppose that* B *is a radical generalized Banach algebra of power series for which* T_h *is an automorphism of* $B^{\#}$ *for all* h *in* $z + B^{(2)}$. *If* g *is a series of order* j *in* B, *then there is a principal automorphism* T_h *for which* $T_h(z^j) = g^j$ *if and only if there is an* h *in* $z + B^{(2)}$ *for which* $h^j = g$.

Proof. $T_h(z^j) = h^j$. So $T_h(z^j) = g$ if and only if $h^j = g$.

The folowing lemma gives several conditions on the series g^j and h^j which are equivalent to the two series being in the same orbit. The hypotheses on B in the following lemma are the same as that in Theorem (9.7).

LEMMA (11.2). *Suppose that* B *is a generalized Banach algebra of power series and that* g *and* h *are series in* $z + S^{(2)}_{-(j-1)}$. *If*

(i) $S_{-(j+2)}(B)$ *is an algebra,*

(ii) z^2D *is a derivation of* $B^{\#}$,

(iii) $\lim_k L_z Q_k = 0$;

then the following are equivalent:

(A) *there is a principal automorphism of* $B^{\#}$ *which maps* g^j *to* h^j.

(B) $T_{\tilde{g}}(B) = T_{\tilde{h}}(B)$.

(C) $\tilde{g} \circ h$ *belongs to* $z + B^{(2)}$.

(D) $\tilde{g}^n - \tilde{h}^n$ *belongs to* B *for* $n = 0,1,2,\ldots j-1$.

Proof. By Theorem (9.7), $T_h T_g^{-1} = T_{\tilde{g}\circ h}$ is the unique principal automorphism of $S_{-(j-1)}$ which maps g^j to h^j. Thus condition (A) merely says that $T_h T_{\tilde{g}}(B) = B$, which is clearly equivalent to condition (B). By Theorem (6.4), $T_{\tilde{g}\circ h}$ is an automorphism of B if and only if $\tilde{g} \circ h$ belongs to

$z + B^{(2)}$, so conditions (A) and (C) are equivalent.

By Theorem (9.7), $T_{\tilde{g}}(B^{(j)}) = T_{\tilde{h}}(B^{(j)}) = B^{(j)}$. Also, $B = \mathrm{span}\{z, z^2, \ldots z^{j-1}\} + B^{(j)}$. Hence condition (B) is equivalent to:

$$\mathrm{span}\{g, g^2, \ldots g^{j-1}\} + B^{(j)} = \mathrm{span}\{h, h^2, \ldots h^{j-1}\} + B^{(j)}.$$

Hence condition (B) is equivalent to:

$$\tilde{g}^n - \tilde{h}^n \in B, \quad \text{for} \quad n = 1, 2, \ldots, j-1.$$

This completes the proof of Lemma (11.2).

Condition (B) above just states that g and h are in the same left cosets of the subgroup $z + B^{(2)}$ in the group $z + S^{(2)}_{-(j-1)}$.

We can now state and discuss a conjecture about the orbits of the automorphism group. The conjecture is consistent with Theorem (11.1) above and with Theorem (11.6) below; but we cannot prove it in the general case. Since this is just a conjecture we do not specify specific hypotheses on B; but presumably the correct hypotheses are approximately those of Lemma (11.2).

CONJECTURE (11.3). *Suppose that* B *is a generalized Banach algebra of power series satisfying suitable hypotheses and that* g *and* h *belong to* $z + S^{(2)}_{-(j-1)}$. *Then the following are equivalent:*

(A) *There is a principal automorphism of* $B^{\#}$ *which maps* g^j *to* h^j.

(B) *There is a principal automorphism of* $B^{\#}$ *which maps* $\tilde{g}^j$ *to* $\tilde{h}^j$.

(C) $g^j - h^j$ *belongs to* $(B)^2$.

Because of condition (11.2) (B), the equivalence of (11.3)(A) and (11.3)(B) is the same as the statement that $z + B^{(2)}$ is a normal subgroup of $z + S^{(2)}_{-(j-1)}$. It is not hard to define a generalized Banach algebra of power series $S_j(B)$ for which $S_{-j}(S_j(B)) = B$ (see [13, pp. 655-656]). So, if conjecture (11.3) is valid, we can use it to find a sequence of normal subgroups of $z + B^{(2)}$.

Theorem (11.1) implies the validity of conjecture (11.3) when either g or h belongs to $z + B^{(2)}$. Theorem (11.6) will prove conjecture (11.3) when both g and h belong to $z + S^{(j+1)}_{-(j-1)}$ (which is the same as both g^j and h^j belonging to $z^j + B^{(2j)}$). In this case, if T_f is the automorphism which takes g^j to h^j, then f belongs to $z + B^{(j+1)}$. If we could find the orbits of T_f in the two cases that f belongs to $z + B^{(j+1)}$ and that f is a polynomial, then we would know the orbits of the full automorphism group, because of Lemma (6.11).

Since many of our arguments involve $(B)^j$, we remark that, if $S_{-1}(B)$ is an algebra, then

$$(B)^j = Bz^{j-1}. \tag{11.4}$$

Formula (11.4) can be proved in the same manner as Lemma (3.9) (or see [14, Lemma (3.4), p. 174]).

The following lemma proves that (11.3)(A) implies (11.3)(C) for automorphisms T_f with f in $z + B^{(j+1)}$.

LEMMA (11.5). *Suppose that* B *is a generalized Banach algebra of power series which satisfies the hypotheses of Lemma (11.2). If* f *belongs to* $z + B^{(j+1)}$ *and* g *belongs to* $B^{(j)}$, *then* $T_f(g) - g$ *belongs to* $(B)^j$.

Proof. Let $f = z + z^2h = z + z^{j+1}h_o$. From formula (6.2) we have

$$T_f(g) - g = z^{j-1} \sum_{k=1}^{\infty} h_o h^{k-1}(z^{2k}D^k g/k!).$$

For all $k \geq 1$, $z^{2k}D^k g$ belongs to $B^{(j+1)}$; also, by Lemma (3.7), L_{h_o} and L_h are bounded operators on $B^{(j+1)}$. Hence each of the summands in the series belongs to B; and the series converges in B, by Lemma (6.5)(A). Therefore $T_f(g)-g$ belongs to $Bz^{j-1} = B^j$; and the proof is complete.

We can now verify Conjecture (11.3), in the promised special case. We again use the same hypothesis as in Lemma (11.2) and Theorem (9.7).

THEOREM (11.6). *Suppose that* B *is a generalized Banach algebra of power series and that* g^j *and* h^j *belong to* $z^j + B^{(2j)}$. *If*

(i) $S_{-(j+2)}$ *is an algebra,*

(ii) z^2D *is a derivation of* $B^{\#}$,

(iii) $\lim_k L_z Q_k = 0$;

then the following are equivalent:

(A) *There is a principal automorphism of* $B^{\#}$ *which maps* g^j *to* h^j.

(B) *There is a principal automorphism of* $B^{\#}$ *which maps* $\tilde{g}^j$ *to* $\tilde{h}^j$.

(C) $g^j - h^j$ *belongs to* B^j.

(D) $\tilde{g}^j - \tilde{h}^j$ *belongs to* B^j.

Proof. By Lemma (11.5), (A) implies (C), and (B) implies (D). Since $\tilde{\tilde{g}} = g$, we can complete the proof by showing that (C) implies (B).

Suppose that k is a fixed integer with $0 < k \leq j$. We prove that:

$$g - h \in B \Leftrightarrow g^k - h^k \in Bz^{k-1} \tag{11.7}$$

Because of Lemma (11.2), this will prove that (C) implies (B), and hence formula (11.7) will imply Theorem (11.6). First, observe that

$$g^k - h^k = (g-h)(g^{k-1} + g^{k-2}h + \dots h^{k-1}), \tag{11.8}$$

and let

$$g^{k-1} + g^{k-2}h + \cdots h^{k-1} = z^{k-1}(k+f).$$

As in the proof of Lemma (3.9), it is easy to show that f belongs to S_{-k}. Dividing formula (11.8) by z^{2k-1}, we obtain

$$(g^k-h^k)/z^{2k-1} = ((g-k)/z^k)(k+f).$$

Since S_{-k} is a radical algebra, we have $(g^k-h^k)/z^{2k-1}$ belongs to $S^{\#}_{-k}$ if and only if $(g-h)/z^k$ does. But this is equivalent to formula (11.7). This completes the proof of Theorem (11.6).

12. WEIGHTED SHIFTS

In Section 1, we observed that the operation of multiplication by z on the space of power series $K^p\langle c_n\rangle^\#$ (see Definitions (1.1) and (1.2)) corresponds to the shift with weights $\{c_{n+1}/c_n\}_{n=o}^\infty$ operating on the ordinary l_p space of the non-negative integers. This power series representation was used to study weighted shifts on l_p by Nikolskii [36] [37]; a similar power series representation of shifts on more general spaces was exploited by Gellar [6] [7] and by the present author [15]. For a given weighted shift L, the above papers studied the invariant and hyperinvariant subspaces and the commuting closed operators and bounded operators. Similar results were obtained without the use of power series by several authors, including Donoghue [3], Nordgren [35], and Lambert [25][23]. In this section we complete the earlier study of the commuting operators of L by finding all closed commuting operators. We also apply the results in the previous sections to study the algebra of bounded operators with commute with L; in particular, we describe the inner automorphisms and generators of this algebra.

We now recall briefly, from [15, Section 2], the construction of the power series representation for an arbitrary shift, L, with non-zero weights. We first need to describe the kind of basis on which L acts as a shift. Suppose that $\{b_n\}_o^\infty$ is a sequence of vectors in the Banach space E and that $\{\beta_n\}_o^\infty$ is a sequence of bounded linear functionals for which $\beta_j(b_k) = \delta_{jk}$. The biorthogonal sequence (b_n,β_n) is said to be a *generalized basis* of E if the functionals $\{\beta_n\}_o^\infty$ separate the points of E; and it is said to be a *Markushevich basis* if, in addition, $\text{span}\{b_n\}_o^\infty$ is dense in E. The generalized basis or Markushevich basis is said to be *normalized* if all $\|b_n\| = 1$.

Now suppose that $\{b_n,\beta_n\}$ is a generalized basis of the Banach space E, that $\{c_n\}_0^\infty$ is a sequence of non-zero scalars, and that L is a bounded weighted shift with weights $\{c_{n+1}/c_n\}_0^\infty$ with respect to the generalized basis (b_n,β_n). The *power series representation determined by* L is the map which associates with x in E the formal power series $\sum_0^\infty \beta_n(x)z^n/c_n$. The space, $B^\#$, of all power series obtained from this representation is a Banach space under the inherited norm. Under this representation, L corresponds to the operator L_z on $B^\#$; c_nb_n corresponds to z^n; and the functional β_n/c_n on E corresponds to the functional π_n on $B^\#$ (see Definition (2.1)). We will call the subspace, B, which consists of the series in $B^\#$ with zero constant term, the *space of power series determined by* L. If the space B is an algebra, it is clearly a generalized Banach algebra of power series. We should point out that, technically speaking, the power series representation and the space of power series is determined not by the shift L, but rather by the generalized basis (b_n,β_n) and the sequence of scalars $\{c_n\}_0^\infty$. ([15, Section 2] has a more precise description of, and additional facts about, the power series representation).

In the rest of this section we suppose that (b_n,β_n) is a fixed normalized generalized basis of a Banach space E, that $\{c_n\}_0^\infty$ is a fixed sequence of non-zero scalars, and that L is the weighted shift on E with weights $\{c_{n+1}/c_n\}_0^\infty$. All the major theorems of this section will be stated for L in terms that do not involve power series. However, the proofs of these theorems will involve the power series representation determined by L; and some secondary results will be stated for algebras of power series. To proceed in this manner we will need to translate properties of the space of power series B, determined by L, into properties of L and E.

First, we define the subspace $E^{(n)}$ of E which corresponds to $B^{(n)}$ (see Definition (2.2) above). So for $n \geq 0$ we define

$$(12.1) \qquad E^{(n)} = \bigcap_{k=o}^{n-1} (\beta_k)^{\perp} = \{x \in E : \beta_o(x) = \beta_1(x) \ldots = \beta_{n-1}(x) = 0\}.$$

Since so many of our results on algebras of power series involve the spaces $S_{-n}(B)$ of Definition (2.4), it will be particularly important to relate these spaces to L and E. It is not difficult to trace through the definitions to show that if L determines the space of power series, B, then the restriction of L to $E^{(n)}$ determines the space of power series $S_{-n}(B)$ (see [15, Th. (6.14)]). However, for our purposes it is far more important to characterize the property that all $S_{-n}(B)$, $n \geq 0$, are radical algebras. We will see, in Theorem (12.3), that the following property of L will be the desired characterization.

DEFINITION (12.2). Suppose that (b_n, β_n) is a normalized generalized basis of the Banach space E, that $\{c_n\}_o^{\infty}$ is a sequence of non-zero scalars, and that L is the shift with weights $\{c_{n+1}/c_n\}_o^{\infty}$. We say that L is *completely cyclic* if:

(i) There is a non-negative integer k for which the range of L^k is contained in the closure of $\text{span}\{b_n\}$.

(ii) For all $n \geq 0$, b_n is a cyclic vector for the algebra of operators on $E^{(n)}$ which commute with the restriction of L to $E^{(n)}$.

Notice that if $\text{span}\{b_n\}$ is dense in E, that is, if (b_n, β_n) is a

Markushevich basis, then condition (i) of Definition (12.2) is automatic.

THEOREM (12.3). *Suppose that* L *is a quasi-nilpotent weighted shift satisfying condition (12.2)(i) and that* B *is the space of power series determined by* L. *Then:*

(A) L *is completely cyclic if and only if all* $S_{-n}(B)$ *are radical algebras.*

(B) If (b_n, β_n) *is a Markushevich basis and* L *is completely cyclic, then, for all* $n \geq 0$, *every operator on* $E^{(n)}$ *which commutes with the restriction of* L *to* $E^{(n)}$ *is a limit, in the uniform operator topology on* $E^{(n)}$, *of polynomials in the restriction of* L *to* $E^{(n)}$.

Proof. If L is a quasi-nilpotent shift satisfying (12.2)(i) and if all $S_{-n}(B)$ are algebras, then it follows from Lemma (3.6) that all $S_{-n}(B)$ are radical algebras. The rest of Theorem (12.3) will follow from Lemma (12.4), below.

Before stating Lemma (12.4), we recall from [15, Def. (2.4)] that a Banach space, B, which is continuously embedded in $C[[z]]$ and which contains only series with zero constant term, is said to be a *generalized Banach module* if B contains all polynomials with zero constant term and if B is closed under multiplication by z. The space of power series determined by a weighted shift is clearly a generalized Banach module. Also, if B is a generalized Banach module, then the spaces $S_{-n}(B)$ of Definition (2.4) above are all generalized Banach modules.

LEMMA (12.4). *Suppose that* B *is a generalized Banach module and that there is a non-negative integer* k *for which* $z^k f$ *is a limit of polynomials for all* f *in* B. *Suppose that* n *is a non-negative integer and let*

$$W_n = \{f \in C[[z]] : fB^{(n)} \subseteq B^{(n)}\}.$$

Then:

(A) $W_n \subseteq S_{-n}(B)^{\#}$.

(B) *If* V *is a bounded operator on* $B^{(n)}$ *which commutes with the restriction of* L_z *to* $B^{(n)}$, *then* $V = L_f$ *for some* f *in* W_n.

(C) *The following are equivalent:*

(1) $W_n = S_{-n}(B)^{\#}$.

(2) z^n *is a cyclic vector for the set of operators on* $B^{(n)}$ *which commute with the restriction of* L_z *to* $B^{(n)}$.

(3) $S_{-n}(B)$ *is an algebra.*

Proof. Part (A) is an immediate consequence of the definitions of S_{-n} and W_n.

We now prove Part (B). Suppose that V is a commuting operator for L_z restricted to $B^{(n)}$. Since Vz^n belongs to $B^{(n)}$, there is a power series f for which

$$Vz^n = fz^n = L_f z^n.$$

Then V and L_f agree on $\text{span}\{z^j\}_{j \geq n}$. Since V is continuous and L_f is closed, V and L_f must agree on the closure of $\text{span}\{z^j\}_{j \geq n}$. Suppose g is an arbitrary series $B^{(n)}$. Then

$$z^k V(g) = V(z^k g) = L_f(z^k g) = z^k L_f(g).$$

This completes the proof of Part (B).

We now prove (C). If $S_{-n}(B)$ is an algebra, then $S_{-n}(B)^{\#} \subseteq W_n$, by Lemma (3.7). If $W_n = S_{-n}^{\#}$, then z^n is cyclic for $\{L_f : f \in W_n\}$ by the definition of S_{-n} (see Definition (2.4)). Finally, suppose (C)(2) holds. If f and g belong to S_{-n}, then fz^n and gz^n belong to $B^{(n)}$. By (C)(2), there is a series h in W_n such that $L_h z^n = hz^n = fz^n$. Hence $f = h$ belongs to W_n, and therefore fgz^n belongs to B. Thus fg belongs to $S_{-n}(B)$; so $S_{-n}(B)$ is an algebra. This completes the proofs of Lemma (12.4) and of Theorem (12.3).

The case $n = 0$ of Lemma (12.4) shows that, if L is a shift satisfying (12.2)(i), then the algebra of operators commuting with L has b_o for a cyclic vector if and only if the space of power series determined by L is an algebra (compare [6, p. 539] and [15, Lemma (4.8) and Theorem (5.6)]). Thus in this case L is a *strictly cyclic shift* in the sense of Lambert [25], [23].

Our earlier paper [15] is largely a study of strictly cyclic and completely cyclic weighted shifts, though this terminology is not used. In particular, [15, Th. (4.1)] shows that if (b_n, β_n) is a Markushevich basis and if L is a completely cyclic shift, then the sequence $\{E^{(n)}\}_1^\infty$ is the completed set of proper non-zero invariant subspaces of L (the corresponding result for generalized bases is [15, Th. (4.4)]). Sufficient conditions for cyclicity or complete cyclicity can be found throughout [15], particularly in Lemma (3.3), Corollary (3.8), and Theorems (4.1), (6.4), (6.7) and (6.9) of [15]. The following simple sufficient condition is part of [15, Th. (4.1)].

LEMMA (12.5). *Suppose that* (b_n, β_n) *is a normalized generalized basis of the Banach space* E, *that* $\{c_n\}_o^\infty$ *is a sequence of non-zero scalars, and that* L *is a weighted shift with weights* $\{c_{n+1}/c_n\}_o^\infty$. *If*

(i) *there is an integer* j *for which* $\{|c_{n+j}|/|c_n|\}_{n=o}^\infty$ *is eventually non-increasing,*

(ii) *there is an integer* m *for which* $\Sigma_{n=o}^\infty \|B_n\| \, |c_{n+m}|/|c_n|$ *converges; then* L *is a completely cyclic quasi-nilpotent shift.*

Proof. Let B be the space of power series determined by L. By hypothesis (i) and [13, Th. (2.10), p. 645], all $S_{-i}(K<c_n>)$ are algebras. By hypothesis (ii), $B \subseteq S_{-m}(K)$. Hence Lemma (12.5) follows from Lemma (3.4) above.

We now study completely cyclic operators and their commuting algebras. Our first two results, Theorems (12.6) and (12.8), assume only that the weighted shift is completely cyclic. These two theorems use Lemma (5.2) and results from Section 3, above, but no other results from the present paper. Theorem (12.6) extends our earlier determination of the closed operators with dense domain that commute with a strictly cyclic shift [15, Ths. (5.1) and (5.6)] (also compare [35] [25] [17, pp. 19-20]).

THEOREM (12.6). *Suppose that* L *is a completely cyclic shift on* E *and that* V *is a non-zero closed operator that commutes with* L. *Then:*

(A) V *is one-one.*

(B) *There is a bounded operator* U *on some* $E^{(m)}$ *for which* U *commutes with the restriction of* L *to* $E^{(m)}$ *and for which* V *or* V^{-1} *is a restriction of* U.

Proof. Let B be the space of power series determined by L. Since we will be working exclusively in the space $B^\#$, instead of E, we can without confusion let V be a non-zero closed operator on $B^\#$ for which

V commutes with L_z.

Let R be the closure of the polynomials in B. Then, for each j, $S_{-j}(R)$ is the closure of the polynomials in the radical algebra $S_{-j}(B)$ (see Definition (2.4)); so all $S_{-j}(R)$ are radical algebras. Also, by Definition (12.2)(i), there is a non-negative integer k for which

$$Bz^k \subseteq R.$$

Let N be the intersection of the null-space of V with $R^{\#}$. If V were not one-one, then N would be a closed non-zero L_z-invariant subspace of $R^{\#}$. In this case N would have to equal $R^{(j)}$ for some $j \geq 0$ (see [15, Th. (4.1)] or [14, Lemma (3.2), p. 173]). Now if f is any series in the domain of V, then fz^{k+j} belongs to $R^{(j)}$; so

$$V(f)z^{k+j} = V(fz^{k+j}) = 0.$$

This contradicts the assumption that V is not a zero operator, and hence proves Part (A) of the theorem.

We now prove Part (B). First we consider the case that the polynomials are dense in B (i.e., that the basis (b_n, β_n) of E is a Markushevich basis). There is a non-negative integer m for which $B^{(m)}$ contains both the domain and range of V, but for which there is some series f of order m in the domain or in the range of V. By Part (A), there is no loss of generality in assuming that f belongs to the domain of V. Hence there is a power series g for which

$$Vf = gf = L_g f.$$

Then g belongs to $S_{-m}(B)^{\#}$, by Lemma (3.8); and L_g is a bounded operator on $B^{(m)}$, by Lemma (3.7). Let

$$M = \{h \in \mathrm{dom}(V) : L_g h = Vh\}.$$

Since V is closed and L_g is continuous, M is a closed subspace of $B^{(m)}$. Moreover M is L_z-invariant and contains the series f which has order m. Hence $M = B^{(m)}$ (see [15, Th. (4.1)] or [14, Lemma (3.2), p. 173]), and V is the restriction of L_g to $B^{(m)}$. This proves Part (B) in the special case that the polynomials are dense in B.

Returning to the general case, we define the closed operator W as the restriction of V to the space of f in $R^{\#}$ for which Vf also belongs to $R^{\#}$. Then, by the above special case, there is a series g in some $S_{-m}(R)^{\#} \subseteq S_{-m}(B)^{\#}$ for which either W or W^{-1} coincides with L_g on $R^{(m)}$; for definiteness assume that it is W which coincides with L_g. Since $B^{\#}z^{k+m} \subseteq R^{(m)}$, V and L_g also coincide, on the domain of V. This completes the proof of Theorem (12.6).

We remark that in the special case that the generalized basis (b_n, β_n) of E is actually a Markushevich basis, then the domain of V is some $E^{(m)}$; and, by Theorem (12.3)(B), V is a uniform limit of polynomials in the restriction of L to $E^{(m)}$. In the general case, the domain of V can be any closed non-zero L-invariant subspace (see [15, Th. (4.4)] for a description of these subspaces).

Suppose that L is a completely cyclic shift, or even just a strictly cyclic shift, on E. Then the space, B, of power series determined by L is an algebra (this is the case $n = 0$ of Lemma (12.3) above). Hence every bounded operator V which commutes with L on E corresponds to a left multiplication operator L_f for some f in $B^{\#}$. We now define the order of V in such a way that it corresponds to the order of f.

DEFINITION (12.7). If L is a strictly cyclic shift with respect to the generalized basis (b_n, β_n) of E, and if V is a non-zero bounded operator which commutes with L, then V is said to have *order* n if

$$\beta_o(Vb_o) = \beta_1(Vb_o) = \cdots = \beta_{n-1}(Vb_o) = 0 \quad \text{and} \quad \beta_n(V(b_o)) \neq 0.$$

If $V = 0$, V has *order* ∞.

Equivalently, V has order n if $V(E) \subseteq E^{(n)}$ but $V(E) \not\subseteq E^{(n-1)}$. If (b_n, β_n) is a Markushevich basis and L is completely cyclic, then the order of V equals the co-dimension of the closure of $V(E)$ in E.

THEOREM (12.8). *Suppose that* L *is a completely cyclic shift on* E *and that* V *is a non-zero bounded operator that commutes with* L. *If* V *has order* $k \geq 1$ *and if* $c_o\beta_k(Vb_o) = \lambda c_k$, *then:*

(A) $\|V^n\|^{1/n}$ *is asymptotic to* $\|(\lambda L^k)^n\|^{1/n}$, *as* n *becomes infinite.*

(B) There are exactly k *bounded operators* W *on* $E^{(k-1)}$ *for which* W^k *is the restriction of* V *to* $E^{(k-1)}$ *and for which* W *commutes with the restriction of* L *to* $E^{(k-1)}$.

Proof. Let B be the space of power series determined by L. Then V corresponds to L_g on $B^{\#}$, with g in $B^{(k)}$ and $\pi_k(g) = \lambda$. Part (A) now follows from Lemma (5.2); and Part (B) follows from Lemmas (3.10) and (3.7).

We are now ready to apply some of our results about automorphisms of generalized Banach algebras of power series to the study of weighted shifts. If L is a strictly cyclic shift on the Banach space E and A is the algebra of bounded operators commuting with L, and if B is the space of power series determined by L, then it follows from Lemma (12.3) that A corresponds to the algebra of left multiplications on $B^{\#}$. Thus there is a natural algebraic and topological isomorphism from A onto $B^{\#}$; so virtually every result in this paper can be translated to a result about the derivations, endomorphisms, or automorphisms of A. In the remainder of this section, we translate a few of the results that seem to have a particularly natural interpretation for the shift L and its commuting algebra A. First we need to define the operators on E that correspond to z^2D and Q_j on $B^{\#}$ (see Definitions (2.1) and (6.3)).

DEFINITION (12.9). Suppose that (b_n, β_n) is a normalized generalized basis of a Banach space E, that $\{c_n\}_o^\infty$ is a sequence of non-zero scalars, and that L is a bounded weighted shift on E with weights $\{c_{n+1}/c_n\}_o^\infty$. We define the closed weighted shift D_L and the sequence of bounded projections $\{Q_j\}_o^\infty$ on E by:

$$\beta_{n+1}(D_L(x)) = (nc_{n+1}/c_n)\beta_n(x) \quad \text{and} \quad \beta_o(D_L(x)) = 0;$$

$$Q_j(x) = x - (\beta_o(x)b_o + \beta_1(x)b_1 \cdots + \beta_{j-1}(x)\, b_{j-1}).$$

Also, we say that x is *convergent* in E if the series $\Sigma_o^\infty \beta_n(x)b_n$ converges in norm to x (compare Definition (2.5)).

Of course if $\{b_n\}_o^\infty$ is a Schauder basis for E, then every vector in E is convergent. Also, it follows from Lemma (6.12) above, that if

D_L is a bounded operator on E and if $D_L x$ is convergent for all x in E, then the sequence $\{\|jLQ_j\|\}_o^\infty$ is bounded. Following [43, p. 69] we say that the sequences $\{x_n\}_o^\infty$ and $\{y_n\}_o^\infty$ in E are *fully equivalent* if there is a linear homeomorphism of E which maps each x_n to y_n. If $\{x_n\}$ is a Schauder basis or a Markushevich basis, then $\{x_n\}$ and $\{y_n\}$ are fully equivalent if and only if they are equivalent bases [31, pp. 62 and 123]. The next result is essentially a translation of Theorem (6.4).

THEOREM (12.10). *Suppose that* L *is a completely cyclic shift on the Banach space* E *with normalized generalized basis* (b_n,β_n), *that* A *is the algebra of operators that commute with* L, *and that* V *is an operator in* A *of order* 1 *and satisfying* $\beta_1(Vb_o) = \beta_1(Lb_o)$. *If*

(i) D_L *is a bounded operator on* E,

(ii) $\lim_j LQ_j = 0$;

then:

(A) *Every algebra automorphism of* A *is an inner automorphism determined by a bounded operator on* E.

(B) *There is an inner automorphism of* A *which maps* L *to* V.

(C) *The sequence* $\{\|V^n\|/\|L^n\|\}$ *is bounded above and is bounded away from zero.*

(D) A *is the algebra of bounded operators that commute with* V. *If* (b_n,β_n) *is a Markushevich basis, then* A *is the uniformly closed algebra, with identity, generated by* V.

(E) *The sequences* $\{L^n b_o\}_o^\infty$ *and* $\{V^n b_o\}_o^\infty$ *are fully equivalent in* E.

Proof. Let B be the space of power series determined by L. Then A corresponds to the space of left multiplications on the algebra $B^\#$, and V corresponds to L_g for some g in $z + B^{(2)}$. It is now clear that every automorphism of A is inner and that there is a one-one correspondence between the inner automorphisms of A and the automorphisms of $B^{\#}$. Also the automorphisms of $B^{\#}$ are all continuous (see [29, Th. 1, p. 2], [30, Th. 10], or [20, Th. 9.1, p. 99]). This proves Part (A). By Theorem (6.4), T_g is an automorphism of $B^{\#}$. Also it is easy to see that $L_g = T_g L_z T_g^{-1}$. This proves (B); and Parts (C) and (D) follow easily from (B).

Finally, the sequences $\{L^n b_o\}$ and $\{V^n b_o\}$ in E correspond to the sequences $\{z^n/c_o\}$ and $\{g^n/c_o\}$, respectively, in $B^{\#}$. Since $T_g(z^n/c_o) = g^n/c_o$, this completes the proof of Theorem (12.10).

Theorem (12.10)(D) can be proved, using Theorem (6.17), when (12.2)(i) and (ii) are replaced with the weaker assumption that there is a non-negative integer k for which $D_L L^k$ is a bounded operator on E and for which $D_L L^k x$ converges in E for all x in E. In the case that E is l_p, or is any space with an unconditional basis, we can improve Theorem (12.10) somewhat.

THEOREM (12.11). *Suppose that* L *is a completely cyclic weighted shift for the unconditional basis* (b_n, β_n) *of the Banach space* E, *and suppose that* V *is a bounded operator which commutes with* L. *If* D_L *is a bounded operator on* E, *then* V *is similar to* L *if and only if* V *has order* 1 *and* $|\beta_1(Vb_o)| = |\beta_1(Lb_o)|$.

Proof. If V is similar to L, then the closure of $V(E)$ must have co-dimension 1 in E, so V has order 1. Also, the sequence $\{\|V^n\|/\|L^n\|\}$ must be bounded above and below; so $|\beta_1(Vb_o)| = |\beta_1(Lb_o)|$ by Theorem (12.10)(C). The rest of Theorem (12.11) follows from applying Theorem (7.6) to the space of power series determined by L.

The following theorem generalizes (12.10)(B) and (C) to operators of order greater than 1. We omit the proof, since it is a direct application of Theorem (9.7).

THEOREM (12.12). *Suppose that* L *is a completely cyclic shift on a Banach space* E *and that* V *is a bounded operator of order* $k \geq 1$ *which commutes with* L *and satisfies* $\beta_k(Vb_o) = \beta_k(L^k b_o)$. *If*

(*i*) D_L *is a bounded operator on* E,

(*ii*) $\lim_j LQ_j = 0$;

then:

(*A*) *There is an inner automorphism, of the algebra of operators on* $E^{(k)}$ *commuting with the restriction of* L *to* $E^{(k)}$, *which maps the restriction of* L^k *onto the restriction of* V.

(*B*) *The sequence* $\{\|V^n\|/\|L^{kn}\|\}_n$ *is bounded above and is bounded away from zero.*

13. SPACES OF ANALYTIC FUNCTIONS

Suppose that $F(x)$ and $G(x)$ are analytic functions of x on some region which is star-shaped about the origin. We define the *convolution* of F and G by the formula:

$$F * G(x) = \int_0^x F(x-t)G(t)dt; \tag{13.1}$$

where the integral is evaluated along a contour, in the region, from 0 to x. Notice that the integration operator has the representation

$$1 * F(x) = \int_0^x F(t)dt. \tag{13.2}$$

If $F(x) = \Sigma_0^\infty a_n x^n$ is analytic in some neighborhood of the origin, we define the formal power series $\hat{F}$, in the indeterminate z, by

$$\hat{F} = \Sigma_1^\infty a_{n-1}(n-1)!z^n = \Sigma_1^\infty F^{(n-1)}(0)z^n. \tag{13.3}$$

If A is a collection of analytic functions, then $\hat{A}$ will be the set of all $\hat{F}$ for which $F(x)$ belongs to A.

In this section we use the map $F \to \hat{F}$ of Formula (13.3) to study convolution and integration in certain spaces of analytic functions on the unit disc. These spaces, called analytic spaces in Definition (13.4) below, include in particular all H^p spaces. In Theorem (13.10), we show that all analytic spaces are radical algebras, under convolution, and we show that if A is an analytic space then all $S_{-j}(\hat{A})$ are radical algebras (see

Definition (2.4) above). This enables us to use the results of Part II of the present paper and of [13], [14] and [15] to study analytic spaces. We describe the ideals and integration-invariant subspaces of analytic spaces in Theorems (13.12), (13.13), and (13.14). In Theorems (13.19), (13.20), and (13.21), we study Volterra integral equations in analytic spaces, and the relation between convolution and differentiation in analytic spaces. In Theorems (13.24) and (13.25), and in Corollary (13.26), we describe the operators that commute with integration or differentiation on analytic spaces. Finally, in Theorems (13.27), (13.30), and (13.31), we describe the derivations, automorphisms, and generators of analytic spaces, as algebras under convolution.

Though the possibility of using the power series representation of formula (13.3) to study convolutions of analytic functions is already clear in Shilov [41, pp. 103 and 114], this representation has, to the best of my knowledge, been used to study convolutions only in [8] and [11].

Though the major theme of this section is the usefulness of the representation $F \to \hat{F}$ of formula (13.3) to study convolutions by means of formal power series, the representation is also useful because it provides us with a collection of naturally occurring radical Banach algebras of power series, and hence sheds considerable light on the possible structure of arbitrary Banach algebras of power series. Indeed, we were led to discover Theorem (6.4) above, precisely because $(H^1)^\wedge$ did not satisfy the hypotheses of Theorem (6.1). It seems quite likely that a study of convolution algebras of analytic functions on regions other than discs would be particularly useful in illuminating the structure of Banach algebras of power series.

We are now ready to give a formal definition of the spaces we will study in this section. Throughout this section $F'(x)$ and $F^{(m)}(x)$ will denote the first and m'th derivatives, respectively, of the analytic function $F(x)$.

DEFINITION (13.4). If A is a vector space of functions analytic on the open unit disc and if A contains all polynomials, then A is an *analytic space* if it satisfies:

(i) F belongs to A whenever F' belongs to A.

(ii) There is a non-negative integer m for which F belongs to A whenever $F^{(m)}$ has absolutely converging Taylor series; and there is a non-negative integer k for which F has absolutely converging Taylor series whenever $F^{(k)}$ belongs to A.

(iii) If $F(x)$ belongs to A, so does $xF(x)$.

A is a *generalized Banach analytic space* if it is an analytic space and a Banach space in which the functionals $F \to F^{(n)}(0)$ are all continuous. A is a *Banach analytic space* if it is a generalized Banach analytic space in which the polynomials are dense.

It is clear that the spaces of analytic functions with l_p summable Taylor series coefficients are generalized Banach analytic spaces. More generally, if A is any subspace of H^1 and if A contains all functions with absolutely converging Taylor series, then A satisfies (13.4)(i) and (ii) (because $F(x)$ belongs to H^1 whenever $F'(x)$ has absolutely convergent Taylor series [5, pp. 48-49]); and so A is an analytic space if it is closed under multiplication by x. A similar argument works if $p < 1$, $q = p/(1-p)$, and $H^q \subseteq A \subseteq H^p$ [5, Th. 5.12, p. 88]. Also, it is clear that the space of derivatives and the space of anti-derivatives of the functions in an analytic space A are themselves both analytic spaces. We should point out that we will not need Property (13.4)(iii) until we begin discussing derivations and automorphisms of analytic spaces, in Theorem (13.27) below.

Since our study of analytic spaces under convolution and integration will be based on the formal power series representation $F \to \hat{F}$, of formula (13.3); we collect a few easily verified properties of this representation.

(13.5) $(F * G)^\wedge = \hat{F}\hat{G}$.

(13.6) $(1 * F)^\wedge = z\hat{F}$.

(13.7) If $F(0) = F'(0) \ldots = F^{(m-1)}(0) = 0$ then $(F^{(m)})^\wedge = \hat{F}/z^m$.

(13.8) $(xF(x))^\wedge = z^2 D(\hat{F})$.

(13.9) $(x^n)^\wedge = n!z^{n+1}$; $(x^{n-1}/(n-1)!)^\wedge = z^n$

Formula (13.5) states that the map $F \to \hat{F}$ is an algebra isomorphism from the algebra, under convolution, of analytic functions on the disc, or on some other star-shaped region, into the algebra of formal power series with zero constant term. The map $F \to \hat{F}$ is defined in purely formal terms in formula (13.3). Though this formal relationship is completely adequate for our purposes, we should point out that there are a variety of more concrete interpretations of formula (13.3) which explain why the map $F \to \hat{F}$ is an algebra isomorphism. Shilov [41, p. 114] interprets the isomorphism of formula (13.3) as a special case of Lorch's abstract characterization of Banach algebras of power series [41, p. 103] [26] [38, pp. 317-318]. In [11] we interpreted the inverse map $\hat{F} \to F$ as a special case of the operational calculus for quasi-nilpotent operators (see [13], [14], and Section 14 below). The map $F \to \hat{F}$ can also be described through the use of Laplace transforms [8, p. 235] [11, p. 416].

We now show that an analytic space A is always a radical algebra, and that the spaces of power series $S_{-j}(\hat{A})$ are also radical algebras (see Definition (2.4)).

THEOREM (13.10). *If* A *is an analytic space, then* A, $\hat{A}$, *and all* $S_{-j}(\hat{A})$ *are radical algebras. If* A *is a (generalized) Banach analytic space, then* $\hat{A}$ *is a (generalized) Banach algebra of power series.*

Proof. Since the map $F \to \hat{F}$ is an algebra isomorphism, by formula (13.5), A will be a radical algebra under convolution if and only if $\hat{A}$ is a radical algebra of formal power series. Let m and k be the fixed integers in Definition (13.4)(ii). Then from formula (13.3) and Definition (1.2) we have

$$\hat{A} \supseteq K\langle 1/(n-k-1)!\rangle;$$

and from formula (13.6) and Definition (2.4) we have

$$S_{-(m+k)}(K) \supseteq \hat{A}\ .$$

It follows from [13, Th. (2.10), p. 645], or from an easy direct calculation, that all $S_{-j}(K\langle 1/(n-k-1)!\rangle)$ are radical algebras. Therefore $\hat{A}$ is an algebra by [13, Lemma (3.6), p. 648]; and is a radical algebra by [13, Lemma (3.8), p. 649].

If A_j is the analytic space of j'th derivatives of functions in A, then it follows from formula (13.6) and Definition (2.4) that

$$\hat{A}_j = S_{-j}(\hat{A});$$

so $S_{-j}(\hat{A})$ is a radical algebra. Finally if A is a (generalized) Banach analytic space, then $\hat{A}$ is a (generalized) Banach algebra of power series, by Definition (1.1). This completes the proof.

Since all $S_{-j}(\hat{A})$ are radical algebras, we can apply results from Section 3 of the present paper, and from [13], [14], and [15], to describe the ideals and integration-invariant subspaces of analytic spaces, and to study convolution multiplication and convolution integral equations. First it will be convenient to define the set $A^{(m)}$ for which $(A^{(m)})\hat{} = (\hat{A})^{(m)}$ (see Definition (2.2)).

DEFINITION (13.11). If A is an analytic space and m is a positive integer, than $A^{(m)}$ is the set of all functions in A which have a zero of order at least $m-1$ at the origin.

The next three theorems describe the ideals and integration-invariant subspaces of analytic spaces. By an *integration-invariant space* we mean a space closed under the integration operator of formula (13.2). It follows from formula (13.6) that J is integration-invariant if and only if $z\hat{J} \subseteq \hat{J}$.

THEOREM (13.12). *Suppose that* J *is a non-zero subspace of the analytic space* A.

(A) *If* J *is an ideal, then there is an* $n \geq 0$ *for which the function* x^n *belongs to* J.

(B) *If* A *is a Banach analytic space and* J *is a closed ideal or a closed integration-invariant subspace, then* $J = A^{(m)}$ *for some* $m \geq 1$.

Proof. Since all $S_{-j}(\hat{A})$ are radical algebras, every ideal in $\hat{A}$ contains a power of the indeterminate z, by [13, Lemma (2.7), p. 644];

this proves Part (A). If A is a Banach analytic space, then $\hat{A}$ is a Banach algebra of power series; so Part (B) follows from [14, Lemma (3.2), p. 173] or [15, Th. (4.1)].

The next theorem gives the version of Theorem (13.12)(B) which is valid for generalized Banach analytic spaces. We omit the proof, which is a straightforward translation of [15, Th. (4.4)].

THEOREM (13.13). *Suppose that* A *is a generalized Banach analytic space, and that* H *is the closure of the polynomials in* A. *If* J *is a closed non-zero subspace of* A, *then the following are equivalent.*

(A) J *is invariant under the integration operator of formula (13.2).*

(B) J *is an ideal of* A.

(C) There is a positive integer m *for which* $A^{(m)} \supseteq J \supseteq H^{(m+1)}$.

Our final result gives somewhat more information on ideals in analytic spaces than Theorem (13.12)(A).

THEOREM (13.14). *Suppose that* A *is an analytic space and that* J *is an integration-invariant non-zero subspace of* A. *Then* J *is an ideal if and only if there is a positive integer* m *for which* $J \subseteq A^{(m)}$ *and for which* J *contains all* m*-fold integrals of functions in* $A^{(m)}$.

Proof. The condition in the theorem translates to the following formula for spaces of power series:

$$\hat{A}^{(m)} \supseteq \hat{J} \supseteq \hat{A}^{(m)} z^m. \tag{13.15}$$

Suppose $\hat{J}$ satisfies formula (13.15). Since J is integration invariant, $z\hat{J} \subseteq \hat{J}$. So $\hat{J}$ will be an ideal in $\hat{A}$ provided $(\hat{A}^{(m)})\hat{J} \subseteq \hat{J}$. But $(\hat{A}^{(m)})\hat{J} \subseteq (\hat{A}^{(m)})^2 = \hat{A}^{(m)} z^m$, by [14, Lemma (3.4), p. 174]. So $\hat{J}$ is an ideal of $\hat{A}$; and hence J is an ideal of A.

Conversely, choose m such that $J \subseteq A^{(m)}$ and J contains a function $F(x)$ which has a zero of order $m-1$ at the origin. Then $\hat{J} \supseteq \hat{A}^{(m)}\hat{F} = \hat{A}^{(m)}z^m$, by [14, Lemma (3.5), p. 175]. Thus J satisfies formula (13.15); and the proof is complete.

Suppose that $F(x)$ and $H(x)$ are analytic functions on the unit disc, and consider the following three Volterra integral equations for the unknown function $X(x)$.

(13.16) $$X - F * X = H$$

(13.17) $$F * X = H$$

(13.18) $$X^{*n} = X * X \cdots *X = H.$$

In [11] we discussed the above three equations, giving a method of solution based on the map $F \to \hat{F}$ of formula (13.3). In the space of all functions analytic on the disc, Equation (13.16) has a unique solution. Also, Equation (13.17) has a solution, which is unique, if and only if $H(x)$ has a zero at the origin of order at least one more than the order of $F(x)$ at the origin. Similarly, Equation (13.18) has an analytic solution if and only if $H(x)$ has a zero at the origin of order one less than a multiple of n; moreover, in this case, there are precisely n analytic solutions (see [11] for the details).

We will examine the solutions of Equations (13.16), (13.17), and (13.18) when the functions $F(x)$ and $H(x)$ belong to, or are related to, some fixed analytic space. We will also examine the convolution products of functions related to an analytic space. These problems are much more difficult for a fixed analytic space than for the space of all analytic functions on the unit disc, because analytic spaces are *never* closed under differentiation (see Definition (13.4)(ii) above).

THEOREM (13.19). *If* F *and* H *belong to the analytic space* A, *then there is a unique* G *in* A *satisfying*

$$G - F * G = H.$$

Proof. Let $A^{\#}$ be the algebra formed by the adjunction of a formal identity u to A [38, p. 2]. Since A is a radical algebra under convolution, u-F is invertible in $A^{\#}$ [38, pp. 17 and 55]. So $G = (u-F)^{-1} * H$ belongs to A.

THEOREM (13.20). *Suppose that* A *is an analytic space, that* F *belongs to* A, *and that* H *is an analytic function on the unit disc. If* F(x) *has a zero of order* $m - 1 \geq 0$ *at the origin and if* H(x) *has a zero of order at least* m *at the origin, then there is a unique analytic function* G *for which*

$$F * G = H.$$

Moreover:

(A) G *is the m'th derivative of a function in* A *if and only if* H *belongs to* A.

(B) G *belongs to* $A^{(m)}$ *if and only if* H *is the* m*-fold integral, using the integral operator of formula (13.2), of a function in* $A^{(m)}$.

Proof. The existence and uniqueness of G is well-known (see [11], for instance). Using formula (13.5) we can replace the equation $F * G = H$ by

$$\hat{F}\hat{G} = \hat{H} .$$

$\hat{F}$ is a series of order m in $\hat{A}$ (see Definition (2.2)). Also, by formula (13.6) and Definition (2.4), G is the m'th derivative of a function in A if and only if $\hat{G}$ belongs to $S_{-m}(\hat{A})$. Part (A) now follows directly from Lemma (3.8) above.

We now prove Part (B). First notice that G belongs to $A^{(m)}$ if and only if $\hat{G}$ belongs to $\hat{A}^{(m)}$. Also, H satisfies the condition of Part (B) if and only if $\hat{H}$ belongs to $\hat{A}^{(m)}z^m$ (see formula (13.6) above). Part (B) now follows from the relation

$$\hat{A}^{(m)}\hat{F} = \hat{A}^{(m)}z^m,$$

which is a special case of [14, Lemma (3.5), p. 175].

If we are willing to allow solutions of Equation (13.17) in $A^{\#}$, instead of just in A, we can weaken the hypothesis of Theorem (13.20)(A) to require only that $H(x)$ have a zero of order at least m at the origin. Then, we can find a function G which is the m'th derivative of a function in A, and we can find a scalar λ, such that

$$\lambda G + F * G = H.$$

The proof is the same as the proof of Theorem (13.20)(A).

THEOREM (13.21). *Suppose that* A *is an analytic space, that* k, m, *and* n *are non-negative integers with* $k \geq n$, *and that* $F_1, F_2, \ldots F_k$ *and* H *are analytic functions on the unit disc. Suppose also that each* F_i *has a zero of order at least* m-1 *at the origin and that* H *has a zero of order exactly* mn-1 *at the origin, and let* G *be one of the* n *analytic functions which satisfies*

$$G^{*n} = G * G \cdots *G = H.$$

(A) If H *belongs to* A, *then* G *is the* (n-1)m'*th derivative of a function in* A. *Conversely, if* $F_1, F_2, \ldots, F_k$ *are* (n-1)m'*th derivatives of functions in* A, *then* $F_1 * F_2 \cdots * F_k$ *belongs to* A.

(B) If the (n-1)m'*th derivative of* H *belongs to* A, *then* G *belongs to* A. *Conversely, if* $F_1, F_2, \ldots, F_n$ *belong to* A, *then the* (n-1)m'*th derivative of* $F_1 * F_2 \cdots * F_n$ *belongs to* A.

Proof. We first prove Part (A). Translating to power series by using formulas (13.3) and (13.5), we have $\hat{G}^n = \hat{H}$ belongs to $\hat{A}$. So $\hat{G}$ belongs to $S_{-m(n-1)}(\hat{A})$, by Lemma (3.10); and hence G is the (n-1)m'th derivative of a function in A, by formula (13.6). Similarly $(F_1 * F_2 \cdots * F_n)^\wedge = \hat{F}_1 \hat{F}_2 \cdots \hat{F}_k$ belongs to $(S_{-m(n-1)}(\hat{A}))^k$, which is a subset of $\hat{A}$, by Lemma (3.9). This completes the proof of Part (A).

Part (B) follows from applying Part (A) to the analytic space of functions whose (n-1)m'th derivative belongs to A. This completes the proof of Theorem (13.21).

A variety of corollaries can readily be obtained by combining Theorems (13.20) and (13.21) above with known facts about derivatives of functions in H^p spaces [5, Chapters 3 and 5] [46]. The following two corollaries illustrate how this can be done.

COROLLARY (13.22). *If* $F(x)$ *and* $G(x)$ *belong to* H^1, *then* $F * G$ *has absolutely converging Taylor series and absolutely continuous boundary values. Conversely if* $K(x)$ *is an analytic function for which* $K(0) \neq 0$, *then* K *belongs to* H^1 *if and only if* $K * K$ *has absolutely continuous boundary values.*

Proof. By Theorem (13.20), $(F * G)'(x)$ belongs to H^1, so $F * G$ has absolutely continuous boundary values, by [5, Th. 3.11, p. 42], and has absolutely converging Taylor series by [5, pp. 48-49]. Conversely, if $K * K$ has absolutely continuous boundary values, then $K * K$ is the derivative of an H^1 function, by [5, Th. 3.11, p. 42]; and so K belongs to H^1, by Theorem (13.21) above. This completes the proof.

COROLLARY (13.23). *Suppose that* F_1, F_2, F_3, *and* F_4 *are univalent analytic functions on the disc. Then*

(A) $F_1 * F_2$ *belongs to* H^p *for all* $p < 1$.

(B) $F_1 * F_2 * F_3$ *belongs to* H^p *for all* $p < \infty$.

(C) $F_1 * F_2 * F_3 * F_4$ *has absolutely converging Taylor series.*

Proof. All F_i belong to H^p for $p < \frac{1}{2}$, by [5, Th. 3.16, p. 50]. Hence, by Theorem (13.21) above, $(F_1 * F_2)'$, $(F_1 * F_2 * F_3)''$, and $(F_1 * F_2 * F_3 * F_4)'''$ all belong to H^p, for $p < \frac{1}{2}$. Corollary (13.23) now follows from [5, pp. 48-49] and [5, Th. 5.12, p. 88].

We now determine the commuting operators for integration and differentiation on analytic spaces. We first consider closed operators on generalized Banach analytic spaces.

THEOREM (13.24). *Suppose that* V *is a closed operator that commutes with the integration operator of formula (13.2) on some generalized Banach analytic space* A. *Then:*

(A) V *is one-one and either* V *or* V^{-1} *is a bounded operator on its domain.*

(B) If V *is a bounded operator on its domain, then there is a function* G *in* A, *a non-negative integer* m, *and a scalar* λ, *such that*

$$V(F) = \lambda F + G^{(m)} * F,$$

for all F *in the domain of* V.

Proof. Part (A) follows from applying Theorems (12.6) and (12.3)(A) to the integration operator. If V is bounded, it follows from Theorem (12.6) that V is the restriction of a bounded operator on $A^{(m)}$ which commutes with integration on $A^{(m)}$. The operator has the form $F \to \lambda F + G^{(m)} * F$, by Lemma (12.4). This completes the proof of Theorem (13.24).

THEOREM (13.25). *If* A *is an analytic space, if* V *is a linear operator, defined throughout* A, *and if* V *commutes with integration on* A, *then there is a function* G *in* A *and a scalar* λ *for which*

$$V(F) = \lambda F + G' * F.$$

Proof. Let $\hat{V}$ be the operator on $\hat{A}$ which corresponds to V on A.

As in the proof of Theorem (13.10), there is a Banach algebra of power series $K = K\langle 1/(n-k-1)!\rangle$ and an integer j for which

$$K \subseteq \hat{A} \subseteq S_{-j}(K).$$

Since we can replace $\hat{V}$ by $L_{z^j}V$ if necessary, there is no loss of generality in assuming that $\hat{V}$ maps K into itself. In this case the restriction of $\hat{V}$ to K is a bounded linear operator, by [20, Th. 9.1, p. 99]. So, by Lemma (12.11) above, there is an h in $S_{-1}(K)^{\#} \subseteq S_{-1}(\hat{A})^{\#}$ for which $\hat{V}(\hat{F}) = h\hat{F}$, for all $\hat{F}$ in K. If $\hat{F}$ belongs to $\hat{A}$, then $z^j\hat{F}$ belongs to K; so

$$z^j\hat{V}(\hat{F}) = \hat{V}(z^j\hat{F}) = z^j h\hat{F}.$$

Hence $\hat{V} = L_h$. Finally there is, by formula (13.6), a scalar λ and a function G in A for which

$$L_h(\hat{F}) = (\lambda F + G' * F)\hat{}$$

for all F in A. This completes the proof.

COROLLARY (13.26). *If* A *is an analytic space and* V *is a linear operator from* A *to* $A^{(2)}$ *which commutes with differentiation on* A, *then there is a function* G *in* A *for which* $V(F) = G' * F$ *for all* F *in* A.

Proof. V commutes with the integration operator of formula (13.2), so $V(F) = \lambda F + G' * F$, by Theorem (13.25). But $\lambda = 0$, since $V(F)$ must vanish at the origin. This completes the proof.

If A is an analytic space, then it is closed under multiplication by the independent variable x (see Definition (13.4)(iii)). Hence, by formula (13.8), z^2D is a derivation of $\hat{A}$. It is quite easy to use the results of Section 4 to find all derivations of A. We may use Theorem (4.1) in our discussion, because the proof of Theorem (4.1) does not require that the algebra of power series be a Banach algebra; but unfortunately we cannot use Corollaries (4.2) or (4.3) unless A is a generalized Banach analytic space.

THEOREM (13.27). *If* A *is an analytic space, then the most general derivation* ∂ *of* A *is a map of the form*

$$\partial(F) = \lambda xF(x) + G'' * (xF(x)), \tag{13.28}$$

where λ *is a scalar and* G *belongs to* A.

Proof. We first show that the map given by formula (13.28) is a derivation. Under the power series representation of formula (13.3), formula (13.28) becomes

$$\partial(F)\hat{} = ((\lambda + (G'')\hat{})z^2)D\hat{F}$$

But $(\lambda + (G'')\hat{})z^2$ is, by formula (13.6), the general form of a series in $\hat{A}^{(2)}$. So ∂ is a derivation, by Theorem (4.1).

Conversely suppose that ∂ is a derivation of A, and let $\hat{\partial}$ be the corresponding derivation of $\hat{A}$. We first show that $\hat{\partial} = gD$ for some g in $\hat{A}$, and then we show that g actually belongs to $\hat{A}^{(2)}$.

As in the proof of Theorem (13.25), we can find a Banach algebra of power series $K = K\langle 1/(n-k-1)!\rangle$ and an integer j for which

$$K \subseteq \hat{A} \subseteq S_{-j}(K).$$

Hence to show that $\hat{\partial} = gD$, we can assume that $\hat{\partial}$ maps K into itself (if not, replace $\hat{\partial}$ by $L_{z^j}\hat{\partial}$). Then by Theorem (2.7), there is a g in $K \subseteq \hat{A}$ for which $\hat{\partial}(f) = gD(f)$ for all f in K. If F belongs to A, then $\hat{F}z^j$ belongs to K and hence

$$\begin{aligned}\hat{F}\,\hat{\partial}(z^j) + \hat{\partial}(\hat{F})z^j &= \hat{\partial}(\hat{F}z^j) = gD(\hat{F}z^j)\\ &= \hat{F}\,gD(z^j) + gD(\hat{F})z^j.\end{aligned}$$

Hence $\hat{\partial} = gD$ for some g in $\hat{A}$.

To complete the proof, we must show that if gD is a derivation of $\hat{A}$, then g cannot have order 1. If g did have order 1, then zD would be a derivation of $\hat{A}$, by Theorem (4.1). We now show that this leads to a contradiction. The operator zD on $\hat{A}$ corresponds to the operator on A which maps $F(x)$ to $(xF)'$, because of formulas (13.7) and (13.8). So if zD maps $\hat{A}$ into itself, then A is closed under differentiation. This contradicts Definition (13.4)(ii), so every derivation of $\hat{A}$ must be of the form gD for some g in $\hat{A}^{(2)}$. Thus every derivation of A has the form given in formula (13.28); so the proof is complete.

We conclude this section by discussing the automorphisms of generalized Banach analytic spaces. The basic question we need to consider is: when do generalized Banach analytic spaces satisfy the hypotheses of the theorems in Section 6? One case is immediate. If A is a Banach analytic space in which the sequence of functions $\{x^n\}_0^\infty$ is a Schauder basis (for instance, if $A = H^p$ for $1 < p < \infty$ [43, Example 11.1, p. 342]), then $z^2D(\hat{F})$ is a convergent formal power series in $\hat{A}$, for all F in A. Hence in this case A will satisfy the hypotheses of Theorem (6.1) and Lemma (6.12).

For many important Banach analytic spaces, such as H^1, H^∞, and the space of analytic functions with continuous boundary values, the sequence $\{x^n\}_o^\infty$ is not a Schauder basis, and hypothesis (ii) of Theorem (6.1) does not hold. Fortunately, in many cases we can still verify the hypotheses of Theorem (6.4). In the rest of this section we identify the space l_p with the space of analytic functions of the unit disc which have their sequence of Taylor coefficients in l_p.

LEMMA (13.29). *Suppose that* A *is a generalized Banach analytic space. If either*

(i) $H^p \subseteq A \subseteq H^1$, *for some* $p < \infty$;

or

(ii) $l_1 \subseteq A \subseteq H^q$, *for some* $q > 1$;

then $\hat{A}$ *satisfies the hypotheses of Theorem (6.4).*

Proof. We need to show that the sequence of operators $\{L_z Q_j\}_j$ approaches zero in the uniform operator topology on $\hat{A}$. To do this we will show that $\lim_j L_z Q_j = 0$ in the operator norm on the space of maps from $\hat{H}^1$ to $\hat{H}^p$ and in the operator norm on the space of maps from $\hat{H}^q$ to $\hat{l}_1 = K\langle 1/(n-1)!\rangle$, where $1 < p, q < \infty$. Define V_j on the space of analytic functions so that the map $F(x) \to xV_j(F)$ corresponds to $L_z Q_j$ on $C[[z]]$. Since multiplication by x is an isometry for each H^p and l_p, it will be sufficient to show that $\lim\|V_j\| = 0$ when the V_j are considered as operators from H^1 to H^p, and as operators from H^q to l_1.

From formulas (13.3) and (13.6), we see that

$$V_j(\Sigma_o^\infty a_n x^n) = \Sigma_{j-1}^\infty a_n x^n/(n+1)$$

so V_j is the multiplier given by the sequence $\{\lambda_n\}$ satisfying

$$\lambda_n = 0 \qquad n < j - 1$$

$$\lambda_n = 1/(n+1) \quad n \geq j - 1.$$

Also, we define the analytic function

$$F_j(x) = \Sigma \lambda_n x^n.$$

First we consider V_j as a multiplier operator from H^1 to H^p, $p < \infty$; and we let p' be the Hölder conjugate of p (i.e., $(1/p) + (1/p') = 1$). It will be enough to restrict ourselves to $p \geq 2$; since decreasing p decreases the norm of V_j. H^p acts as a space of multipliers from H^1 to H^p and the multiplier norm is no greater than the H^p norm. (This follows, for instance, from the fact that the circle-group convolution of L^1 with L^p is contained in L^p). Hence $\|V_j\|$ is less than or equal to the H^p norm of $F_j(x)$, which in turn is less than or equal to the $l_{p'}$ norm of $F_j(x)$, by the Hausdorff-Young Theorem [5, Th. 6.1, p. 94]. This completes the proof of Lemma (13.29) in the case that A satisfies Condition (i).

We now consider V_j as a multiplier operator from H^q to l_1; and we let q' be the Hölder conjugate of q. It will be enough to consider $q \geq 2$, so, by the Hausdorff-Young Theorem [5, Th. (6.1), p. 94], $H^q \subseteq l_{q'}$. Then the norm of V_j as an operator from H^q to l_1 is less than or equal to the l_q norm of $F_j(x)$. This completes the proof of Lemma (13.29).

If A satisfies the hypothesis of Lemma (13.29) above, then $\hat{A}$

satisfies the hypotheses of Theorem (9.7) as well as the hypothesis of Theorem (6.4), since all $S_{-j}(\hat{A})$ are algebras by Theorem (13.10). Hence nearly all the results of Part II of this paper apply to $\hat{A}$; in particular, $\hat{A}$ satisfies the conclusions of Theorem (2.6). The main exceptions are the two results, Lemma (10.5) and Theorem (10.6), that require the sequence $\{\|jL_zQ_j\|\}$ to be bounded; and the special results that assume that z^2D is compact or that (z^n, π_n) is an unconditional basis.

The next theorem applies some of our results about principal automorphisms to algebras satisfying the hypotheses of Lemma (13.29).

THEOREM (13.30). *Suppose that* A *is a generalized Banach analytic space that satisfies the hypotheses of Lemma (13.29).*

(A) If F *belongs to* A *and* $F(0) = 1$, *then there is a continuous automorphism of* A *which maps* 1 *to* F.

(B) If F *belongs to* $A^{(m)}$ *and* $F^{(m-1)}(0) = (m-1)!$, *then there is a continuous automorphism of* $A^{(m)}$ *which maps* x^{m-1} *to* F. *Moreover, this automorphism extends to an automorphism of the analytic space of* (m-1)'st *derivatives of the functions in* A.

Proof. Part (A) follows from Lemma (13.29) and Theorem (6.4); and Part (B) follows from Lemma (13.29) and Theorem (9.7). This completes the proof of the theorem.

For generalized Banach analytic spaces which do not satisfy the hypothesis of Lemma (13.29), we can still obtain some information about automorphisms and generators by applying Theorem (6.15) or Theorem (6.17). The following is an easy application of Theorem (6.17).

THEOREM (13.31). *If* A *is a Banach analytic space and* F(x) *is a function in* A *for which* $F(0) \neq 0$, *then* A *is generated, as a Banach algebra under convolution, by* F.

Proof. By Definition (13.4), there is an integer n, for which $z^n Df$ converges absolutely in $\hat{A}$ for all f in $\hat{A}$. Thus $\hat{A}$ satisfies the hypothesis of Theorem (6.17).

14. OPERATIONAL CALCULUS

Throughout this section, we let T be a fixed quasi-nilpotent, but not nilpotent, operator on a Banach space. If $f = \Sigma_o^\infty \lambda_n z^n$, we define the operator $f(T)$ by

$$f(T)\phi = \Sigma_o^\infty \lambda_n T^n \phi,$$

whenever the series converges weakly (compare [14, p. 182]). Also we denote the series $\Sigma \lambda_n T^n$ by $\overline{f}(T)$ (compare [13, p. 641] or [14, p. 170]). In this section we continue the study, begun in [13] and [14], of the operators $f(T)$ and the map $f \to f(T)$. As in [13] and [14], everything in this section would be completely unchanged by systematically substituting strong convergence for weak convergence; for everywhere-defined operators we could also use uniform or uniform-absolute convergence.

We now establish some convenient notational conventions. We let

$$\|T^n\| = c_n;$$

and we let

$$K = K<c_n>,$$

which is thus the set of all f with zero constant term for which $\overline{f}(T)$ converges absolutely in the uniform operator topology. For $f = \Sigma \lambda_n z^n$, we set

$$\|f\|_K = \Sigma|\lambda_n|c_n ,$$

which is the norm of f in K. Finally we let B be the set of series f with zero constant for which $\bar{f}(T)$ converges in the weak operator topology. Notice that, for all $j \geq 0$,

$$f \in S_{-j}(B) \Leftrightarrow \mathrm{Dom}(f(T)) \supseteq T^j(E);$$

and, in particular, $f(T)$ is an everywhere defined (and hence bounded) operator if and only if f belongs to B.

The operators we study in this section will satisfy the same conditions as the quasi-nilpotent operators studied in [13] and [14]; so we repeat the following definition from [13, Th. (3.1), p. 648] and [14, Def. (4.1), p. 182].

DEFINITION (14.1). The quasi-nilpotent operator T is said to be *amenable* if

(i) $B \subseteq S_{-m}(K)$, for some $m \geq 0$.

(ii) $S_{-j}(K)$ is an algebra, for some $j \geq 0$.

Also, if T is amenable, we define

$$S_{-\infty} = \cup S_{-j}(B) = \cup S_{-j}(K). \tag{14.2}$$

Notice that it follows from the Banach-Steinhaus theorem that

$$K \subseteq B \subseteq K^{\infty}\langle c_n\rangle$$

So (14.1)(i) holds if $\Sigma_n c_{n+m}/c_n$ converges. There are a variety of conditions on $\{c_n\}$ that imply (14.1)(ii); the simplest is the existence of a

$k > 0$ for which $\{c_{n+k}/c_n\}_n$ is eventually non-increasing [13, Th. (2.10), p. 645] (another simple condition is given in [13, Th. (4.7), p. 652]).

The major results of [13], given in [13, Def. (1.2), p. 641-642] and [13, Th. (3.1), p. 646], are: B is an algebra; the map $f \to f(T)$ is an injective algebra homomorphism; and, if f belongs to B and if g has positive radius of convergence, then $g \circ f \in B^{\#}$ and $g \circ f(T) = g(f(T))$. These results are extended, with appropriate modifications, to f belonging to $S_{-\infty}$ in [14, Th. (4.3), pp. 183-187].

In this section we consider $g \circ f$ and $g(f(T))$ when g has zero radius of convergence, and we find bounds on the sequence $\|f(T)^n\|$ for f in B. Notice that for all f,

(14.3) $$\|f(T)\| \leq \|f\|_K.$$

We begin by reducing the problem of studying $g \circ f$ and $g(f(T))$ to the problem of bounding $\{\|f^n\|_K\}_n$.

LEMMA (14.4). *Suppose that* $g = \Sigma_1^\infty \lambda_n z^n$, *that* m *is a positive integer, and that* f *is a series with zero constant term for which* $\Sigma_{n=m}^\infty |\lambda_n| \, \|f^n\|_K$ *converges. If* T *is an amenable quasi-nilpotent operator, then:*

(A) $g(f(T)) \subseteq g \circ f(T)$;

(B) $\mathrm{Dom}\, g(f(T)) = \mathrm{Dom}\, f(T)$;

(C) If f *belongs to* B, *or to some* $S_{-j}(B)$, *so does* $g \circ f$.

Proof. Since $g \circ f$ belongs to $S_{-j}(B)$ for some $j \geq 0$ if and only if $\mathrm{Dom}(g \circ f(T)) \supseteq T^j(E)$, Part (C) will follow from Parts (A) and (B). Also, it follows from the definition of $g(f(T))$ as $\Sigma \lambda_n (f(T))^n$

that $\mathrm{Dom}(g(f(T))) \subseteq \mathrm{Dom}(f(T))$.

Thus, to complete the proof we fix

$$\phi \in \mathrm{Dom}(f(T));$$

and we show that

$$\phi \in \mathrm{Dom}(g(f(T)) \cap \mathrm{Dom}(g^{o}f(T)) \quad \text{and} \quad g(f(T))\phi = g^{o}f(T)\phi.$$

We let $g = p + g'$, where p is a polynomial and g' is a series of order at least m. For each $n > 0$

$$(14.5) \qquad (f(T))^{n} \subseteq f^{n}(T) \qquad \text{and} \qquad \mathrm{Dom}(f(T))^{n} = \mathrm{Dom}\ f(T),$$

by [14, Th. (4.3)(C), p. 184]. Hence

$$\phi \in \mathrm{Dom}\ p(f(T)) \cap \mathrm{Dom}(p^{o}f(T)) \quad \text{and} \quad p(f(T))\phi = p^{o}f(T)\phi.$$

Also, by hypothesis, the series $\Sigma_m^\infty \lambda_n f^n$ must converge in the norm of K, and hence in the Fréchet topology on K; thus $g'^{o}f \in K$ (compare [15, Lemma (3.11)]). It follows from formula (14.3) that $\Sigma_m^\infty \lambda_n f^n(T)$ converges absolutely in the uniform operator norm to $g'^{o}f(T)$. Hence it follows, from formula (14.5), that $g'(f(T))\phi = g'^{o}f(T)\phi$ for ϕ in the domain of $f(T)$. This completes the proof of Lemma (14.4).

The hypothesis of Lemma (14.4) requires, in particular, that there exist an m for which $f^n \in K$ whenever $n \geq m$. By Lemmas (3.9) and (3.10) above, this is equivalent to $f \in S_{-\infty}$ and also to $f^m \in K$. We now combine formula (14.3) with various estimates from Section 10 to obtain bounds on $\{\|f^n(T)\|\}_n$ for $f \in S_{-\infty}$. Of course if $f \in B$, then $f^n(T) = (f(T))^n$; but for general $f \in S_{-\infty}$ we only have $(f(T))^n \subseteq f^n(T)$ [14, Th. (4.3)(C), p. 184].

THEOREM (14.6). *Suppose that* T *is an amenable operator, that* f *belongs to* $S_{-\infty}^{(k)}$, *and that* $|\lambda| > 0$. *If* $|\pi_k(f)| \le |\lambda|$, *then*

(A) $\lim_n (\|f^n(T)\|/\|T^{kn}\|)^{1/n} \le \lim_n (\|f^n\|_K/c_{nk})^{1/n} \le |\lambda|$.

(B) *If* $\{nc_{n+1}/c_n\}$ *is bounded, there is an* $M > 0$ *for which*

$$\|f^n(T)\| \le \|f^n\|_K \le M|\lambda|^n\|T^{nk}\|,$$

for all sufficiently large n.

(C) *If* $\lim(nc_{n+1}/c_n) = 0$ *and if* $\pi_k(f) = \lambda$; *then*

$$\lim_n \|f^n(T) - \lambda^n T^{kn}\|/\|\lambda^n T^{kn}\| = 0$$

and $\|f^n(T)\|$ *is asymptotic to* $\|\lambda^n T^{nk}\|$ *as* n *approaches infinity.*

Proof. Because of formula (14.3), Part (A) follows from Corollary (10.2), Part (B) follows from Theorem (10.3), and Part (C) follows from Theorem (10.7).

The next two corollaries are illustrative of a variety of results that can be obtained from combining Lemma (14.4) with Theorem (14.6). We omit the straightforward proofs. Recall from [13, Def. (4.1), p. 651] that if f is a formal power series and t is a scalar, then

$$f_t = f \circ (tz) = T_{tz} f.$$

COROLLARY (14.7). *Suppose that* T *is an amenable quasi-nilpotent operator, that* $f \in S_{-j}(B)$ *for some* $j \geq 0$, *and that* $g_t \in K$ *for all* $t > 0$. *Then*

(A) $g(f(T)) \subseteq g^{o}f(T)$.

(B) $\text{Dom } g(f(T)) = \text{Dom } f(T) \supseteq T^{j}(E)$.

(C) $g^{o}f \in S_{-j}(B)$.

COROLLARY (14.8). *Suppose that* T *is an amenable quasi-nilpotent operator for which* $\{nc_{n+1}/c_n\}$ *is bounded, that* f *belongs to* B, *and that* g *belongs to* K. *If* $|\pi_1(f)| \leq 1$, *then:*

(A) $g^{o}f$ *belongs to* B.

(B) $g^{o}f(T) = g(f(T))$.

One special case of Corollary (14.8) is worth noting. If $\pi_1(f) = \lambda \neq 0$, and if g is the compositional inverse of f/λ, then $g^{o}(f/\lambda) = g_{(1/\lambda)} \circ f = z$. Thus if f is any series of order one for which $\overline{f}(T)$ converges absolutely, and if T is an amenable operator with $\{nc_{n+1}/c_n\}$ bounded, then T is the sum of an absolutely converging power series in $f(T)$. We can show the weaker result, that T is a uniform limit of polynomials in $f(T)$, under much less stringent hypotheses.

THEOREM (14.9). *Suppose that* T *is an amenable operator and that there is a* $k \geq 0$ *for which the sequence* $\{nc_{n+k+1}/c_n\}_n$ *is bounded. If* f *is a series of order* 1 *for which* $\overline{f}(T)$ *converges in the uniform operator topology, then* T *is a uniform limit of polynomials in* $f(T)$.

Proof. We first suppose that $\overline{f}(T)$ converges absolutely in the uniform operator topology. In this case it follows from Theorem (6.17) that f is a topological generator of K. Hence there is a sequence of polynomials $\{p_n\}$ for which

$$\lim_n \|p_n(f(T)) - T\| \le \lim_n \|p_n(f) - z\|_K = 0$$

We complete the proof, in the general case in which $\overline{f}(T)$ is not assumed to converge absolutely, by showing that for all sufficiently large j, T^j is a limit of polynomials in $f(T)$. This will imply that there is a polynomial p of order 1 for which $p(T)$ is a limit of polynomials in $f(T)$. Without loss of generality, we can assume that $f - z$ is a series of order at least $k + 1$; for we can always find a polynomial in f which satisfies this condition.

By Definition (14.1), there is an $m \ge 0$ for which $B \subseteq S_{-m}(K)$. By Theorem (6.15), T_f is an automorphism of $S_{-m}(K)$. Hence, by Theorem (9.4)(B), T_f is a continuous automorphism of $K^{(m+1)}$. Therefore the sequence $\{f^n\}_{n=m+1}^{\infty}$ generates the subalgebra $K^{(m+1)}$ of the Banach algebra K. So if $j \ge m + 1$, there is a sequence of polynomials $\{q_n\}$ for which

$$\lim_n \|q_n(f(T)) - T^j\| \le \lim_n \|q_n(f) - z^j\|_K = 0.$$

This completes the proof of Theorem (14.9).

Without changing the proof of Theorem (14.9) in any essential way, we could show that if $\overline{f}(T)$ converged in the weak (strong) operator topology, then T is the weak (strong) limit of polynomials in T.

REFERENCES

1. T. M. Apostol, Mathematical Analysis, Addison-Wesley, Reading, Mass., 1957.

2. H. Cartan, Elementary Theory of Analytic Functions of One or Several Complex Variables, English edition, Addison-Wesley, Reading, Mass., 1963.

3. W. F. Donoghue, Jr., The Lattice of Invariant Subspaces of a Completely Continuous Quasi-Nilpotent Transformation, Pacific J. Math., 7 (1957), 1031-1035.

4. N. Dunford and J. T. Schwartz, Linear Operators, Part I, Pure and Applied Mathematics VII, Wiley-Interscience, New York, 1958.

5. P. L. Duren, Theory of H^p Spaces, Pure and Applied Mathematics, 38, Academic Press, New York, 1970.

6. R. Gellar, Operators Commuting with a Weighted Shift, Proc. Amer. Math. Soc., 23 (1969), 538-545.

7. R. Gellar, Cyclic Vectors and Parts of the Spectrum of a Weighted Shift, Trans. Amer. Math. Soc., 146 (1969), 69-85.

8. J. I. Ginsberg and D. J. Newman, Generators of Certain Radical Algebras, J. Approximation Theory, 3 (1970), 229-235.

9. S. Grabiner, Radical Banach Algebras and Formal Power Series, thesis, Harvard, 1967.

10. S. Grabiner, The Nilpotency of Banach Nil Algebras, Proc. Amer. Math. Soc., 21 (1969), 510.

11. S. Grabiner, The Use of Formal Power Series to Solve Finite Convolution Integral Equations, J. Math. Anal. Appl., 30 (1970), 415-419.

12. S. Grabiner, Ranges of Quasi-Nilpotent Operators, Illinois J. Math., 15 (1971), 150-152.

13. S. Grabiner, A Formal Power Series Operational Calculus for Quasi-Nilpotent Operators, Duke Math. J., 38 (1971), 641-658.

14. S. Grabiner, A Formal Power Series Operational Calculus for Quasi-Nilpotent Operators, II, J. Math. Anal. Appl., 43 (1973), 170-192.

15. S. Grabiner, Weighted Shifts and Banach Algebras of Power Series, Amer. J. Math., to appear.

16. S. Grabiner, Ranges of Products of Operators, Canad. J. Math., to appear.

17. D. A. Herrero, Operator Algebras of Finite Strict Multiplicity, Indiana U. Math. J., 22 (1972), 13-24.

18. E. Hewitt and K. A. Ross, Abstract Harmonic Analysis, I., Grund. der Math., 115, Springer-Verlag, Berlin, 1963.

19. N. Jacobson, Structure of Rings, Rev. ed., Amer. Math. Soc. Colloq. Publ., Vol. 37, Amer. Math. Soc., Providence, R.I., 1964.

20. B. E. Johnson, Continuity of Linear Operators Commuting with Continuous Linear Operators, Trans. Amer. Math. Soc., 128 (1967), 88-102.

21. H. Kamowitz and S. Scheinberg, Derivations and Automorphisms of $L^1(0,1)$, Trans. Amer. Math. Soc., 135 (1969), 415-427.

22. H. Kamowitz and S. Scheinberg, The Spectrum of Automorphisms of Banach Algebras, J. Functional Analysis, 4 (1969), 268-276.

23. E. Kerlin and A. Lambert, Strictly Cyclic Shifts on l_p, Acta Sci. Math. (Szeged), 35 (1973), 87-94.

24. J.-L. Lagrange, Sur une Nouvelle Espèce de Calcul Relatif à la Différentiation et à l'Intégration des Quantités Variables, Oeuvres de Lagrange, Volume III, Gauthier-Villars, Paris, 1869. Reprinted from

Nouveaux Mémoires de l'Académie Royale des Sciences et Belles-Lettres de Berlin (1772), 185-221.

25. A. Lambert, Strictly Cyclic Weighted Shifts, Proc. Amer. Math. Soc., 29 (1971), 331-336.

26. E. R. Lorch, The Structure of Normed Abelian Rings, Bull. Amer. Math. Soc., 50 (1944), 447-463.

27. R. J. Loy, Continuity of Derivations on Topological Algebras of Power Series, Bull. Austral. Math. Soc., 1 (1969), 419-424.

28. R. J. Loy, Uniqueness of the Complete Norm Topology and Continuity of Derivations on Banach Algebras, Tôhoku Math. J., 22 (1970), 371-378.

29. R. J. Loy, Uniqueness of the Fréchet Space Topology on Certain Topological Algebras, Bull. Austral. Math. Soc., 4 (1971), 1-7.

30. R. J. Loy, Banach Algebras of Power Series, preprint.

31. J. T. Marti, Introduction to the Theory of Bases, Springer Tracts in Natural Philosophy, 18, Springer-Verlag, New York, 1969.

32. J. K. Miziolek, T. Müldner, and A. Rek, On Topologically Nilpotent Algebras, Studia Math., 43 (1972), 41-50.

33. L. Nachbin, The Haar Integral, Van Nostrand, Princeton, 1965.

34. D. J. Newman, A Radical Algebra without Derivations, Proc. Amer. Math. Soc., 10 (1959), 584-586.

35. E. A. Nordgren, Closed Operators Commuting with a Weighted Shift, Proc. Amer. Math. Soc., 24 (1970), 424-428.

36. N. K. Nikolskii, Basicity and Unicellularity of Weighted Shift Operators, Math. USSR-Izvestija, 2 (1968), 1077-1089.

37. N. K. Nikolskii, Spectral Synthesis for a Shift Operator and Zeros in Certain Classes of Analytic Functions Smooth up to the Boundary,

Soviet Math. Dokl., 11 (1970), 206-209.

38. C. E. Rickart, General Theory of Banach Algebras, Van Nostrand, Princeton, 1960.

39. S. Scheinberg, Power Series in One Variable, J. Math. Anal. Appl., 31 (1970), 321-333.

40. S. Scheinberg, Automorphisms of Commutative Banach Algebras, from R. C. Gunning, ed., Problems in Analysis, A symposium in honor of Salomon Bochner, Princeton University Press, Princeton, 1970.

41. G. Shilov, On Regular Normed Rings, Trudy Math. Inst. Steklov, 21 (1947), 1-118.

42. H. J. Schultz, Systems of Derivations on Topological Algebras of Power Series, J. Austral. Math. Soc., to appear.

43. I. Singer, Bases in Banach Spaces I, Grund. der Math., 154, Springer-Verlag, New York, 1970.

44. P. Szwed, A Scale of a Radical in Commutative B_0-algebras, Bull. Acad. Polon. Sci. Ser. Math. Astronom. Phys., 20 (1972), 745-752.

45. F. Treves, Topological Vector Spaces, Distributions, and Kernels, Pure and Applied Mathematics, 25, Academic Press, New York, 1967.

46. J. Wells, On the Zeros of Functions with Derivatives in H_1 and H_∞, Canad. J. Math., 22 (1970), 342-347.

47. W. Zelasko, Metric Generalizations of Banach Algebras, Dissertationes Math. Rozprawy Mat., 47 (1965).

48. G. Zeller-Meier, Sur les Automorphismes des Algèbres de Banach, C. R. Acad. Sci. Paris Sér. A, 264 (1967), 1131-1132.

Claremont Graduate School, Claremont, California